SALAMANCA
ART AND MONUMENTS

EDILERA

Agradecimientos:

por la facilidad para tomar fotografías y buen trato recibido, por orden alfabético:

Excma. Diputación de Salamanca.
Frailes Capuchinos.
Guías de Antiqua Sedes Castrorvm y Cerro de San Vicente.
Ilustre Cofradía de la Vera Cruz de Salamanca.
Macarro Alcalde, C.
M.M. Clarisas.
M.M. Dominicas.
Museo de Salamanca.
Obispado de Salamanca.
P.P. Dominicos.
Siervas de San José.
Universidad de Salamanca.
Universidad Pontificia de Salamanca.

EDILERA
www.edilera.com
contacto@edilera.com
Zamora, España.

ISBN: 978-84-943610-3-6
Depósito Legal: ZA-026-2015 Y [ZA-91-2022] complemento digital por código QR.
Escrito y maquetado en Zamora, España.
Impreso en Burgos, España. Fabricado en la Unión Europea.

I

N

D

E

X

4 History of the city
prerroman boar ~Ordo Salmanticense
6 Cerro de San Vicente
Early Iron Age cabins
8 Salamanca and St. Cebrian Cave.
Iron Age wall.
Sedes Antiqva Castrorvm
10 Via de la Plata
12 **Roman bridge**
14 Botanical Archeological Park
15 **The Cathedrals**
18 Old Cathedral of St. Mary's of the Seat
History
20 Cockerel tower
22 The interior
24 The cloister, St.Barbera's Chapel.
26 New Cathedral of the Assumption of the Virgin
History
28 Birth Gateway
32 Ramos Gateway
35 The interior
36 El Mariquelo
37 Ieronimus - Ascending the tower
38 Anaya Square
Church of Saint Sebastian
University College of Anaya
40 Anaya Guest House
44 Patio Chico
45 Calixto y Melibea Garden
46 Lis House, Art Noveau and Deco Museum
48 Salamanca Museum
52 **Salamanca University**
History
56 Escuelas Menores
58 The Sky of Salamanca
60 Study Hospital
62 The Façade of Escuelas Mayores
66 The interior
67 The Chapel
68 Paraninfo
70 Fray Luis de León, Classroom
72 The University Stairway
74 The Enigmas
76 The University Library
Liber Canticorvm et Horarvm
78 House-Museum of Miguel de Unamuno
78 Saint Isidore and Saint Pelayus
79 Calle Libreros
80 Conchas House
84 **The Clergy, Pontifical University**
History
The Royal College
87 Scala Coeli
88 The Cloister
90 Paraninfo
92 The Church
94 The Saint Benedict's corner
Saint Benedict's church
Madre de Dios Convent
Maldonado House
95 Solís House
96 Monterrey Palace
98 Church and Convent of the Immaculate C.
102 Chapel of the Third Order of Saint Francis
104 **Archbishop Fonseca University College**
110 Vera Cruz Chapel
111 Saint Ursula's Convent
112 House of the Deaths
114 Saint Mark
116 **Plaza Mayor**
122 Saint Martin of Tours
126 Salina Palace
128 Torre del Aire
129 Torre del Clavero
130 Sancti Spiritus
132 Real Convento de Santa Clara ~Museum
136 Saint Thomas of Canterbury
138 **Convento de San Esteban**
History
141 Façade
143 Porter's lodge
144 Monarchs's Cloister
146 Soto Stairway
147 Sacristy and Old and New Chapters
148 The Church
150 Convent of the Dominican Dueñas
154 Columbus Square
Orellana Palace
Abrantes Tower
Columbus statue
155 Other Monuments of Interest

160 Alphabetic index

Internal front flaps:
Recommendations

Rear flaps internal:
The province of Salamanca

ÍNDICE QR Y VISITA VIRTUAL

BOAR

Alongside the Roman bridge there lies this Pre-Roman sculpture in the nearby archaeological excavations of St.Vincent's hill[6] whose purpose was to serve as an entity to protect the livestock. This acephalous representation today looked like a bull or boar which, over time, has gradually changed its location. Like other monuments in Salamanca, this also appears in one of the most representative works of Spanish literature ***El Lazarillo de Tormes*** in the scene in which the blindman tells then boy that by bringing his head closer to the animal's belly he can hear a great roar and he is pushed strongly against it.

ORDO SALMANTICENSE

M.SA.

IMP CAES
M·AVR·ANTO
NINO AVG
F·L·SEP·SEVE
RI ORDO
SALMANTIC

To Emperor Caesar Marcus Aurelius Antoninus Augustus, son of Lucius Septimius Severus, the *Ordo Salmanticense*.

HISTORY OF THE CITY OF SALAMANCA

The city is situated in a prime geographic location on the route joining the north of the Peninsula to the south, from ancient times to today. Its position on various escarpments on the fertile plain of the River Tormes made it a place which was easy to defend and this aspect, along with the above, gave rise to the birth of the city. The oldest known settlement in the area occupied by the city is located on St.Vincent's hill; this nucleus was gradually extended, reaching the Teso de las Catedrales, and becoming the city of Salmantica in Roman times when it acquired the category of **Civitas** as is borne out by the **Ordo Salmanticensis** kept at the Fine Arts' Museum[48].

With the Barbarian and Moslem invasions the city fell into oblivion and did not undergo any major change as it remained a residual population settlement.

In the early **12th century** Salamanca regained its former splendour; its repopulation as undertaken by order of Alfonso VI who commissioned this to his son-in-law **Raymond of Burgundy**. Franks, toresans and Galicians arrived here to resettle the city and it was they who founded the main Romanesque parish churches, some of which we can still visit such as St.Martin[122] or St.Thomas Becket[136].

In the early 13th century, in **1218**, an event occurred which totally changed the life of the city and its inhabitants until today; on this date the **General Study of Salamanca** or University was founded. Universities and Schools became to appear around it, Military and Religious orders were set up and the city increased its population.

The 15th century was characterised by the confrontations by the main noble families of the city which divided up into two groups around the parish churches, St. Benedict[94] head of the Beniditines and St.Thomas[159], head of the thomesins. The bloodiest moment of this confrontation arrived with the death of the Monroy brothers and the subsequent revenue of María la Brava[159]. The violent episodes reached their conclusion with the mediation of St.John of Sahagún and the action of the Catholic Monarchs who terminated the power of these families; the monarchs intervention resulted in the modification of the urban structure as many palaces and towers were destroyed as a symbol of the power of the monarchy over the nobility.

The **16th century** entailed the height of the splendour of the University and of the city which changed greatly with the construction of new **Plateresque** buildings.

The **17th century**, owing to the prevailing crisis saw the start of the decline which continued until the **19th century**, at the beginning of which the city suffered the **French invasion** and many of its buildings and monuments fell victim to misuse by the Napoleonic troops which devastated and plundered them. The disentailments which followed during the course of this century put paid to the work started by the French invasion and permanently ruined monasteries and convents.

In the **20th century** during the Civil War Salamanca was the headquarters of the national group which is why when the dispute ended there was a major urbanistic remodelling in an attempt to modernize the city.

EL CERRO DE SAN VICENTE[76]

EARLY IRON AGE CABINS

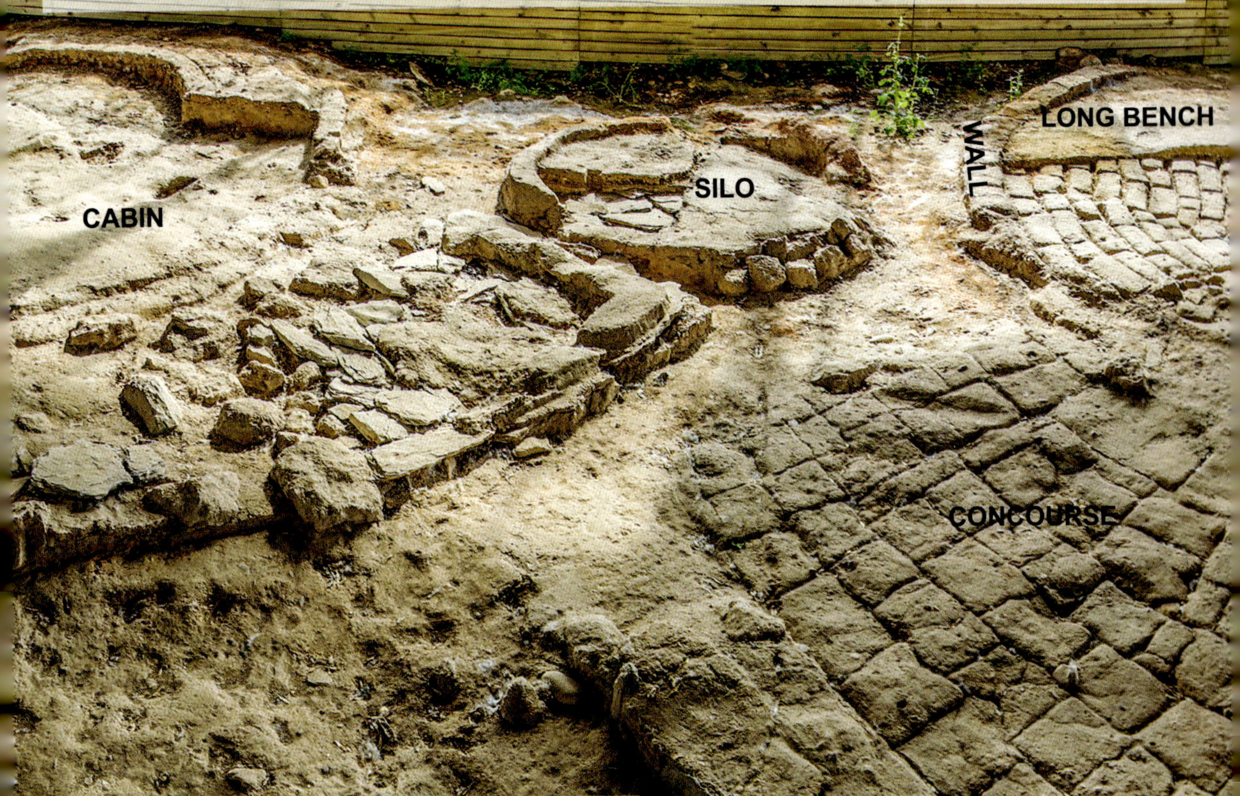

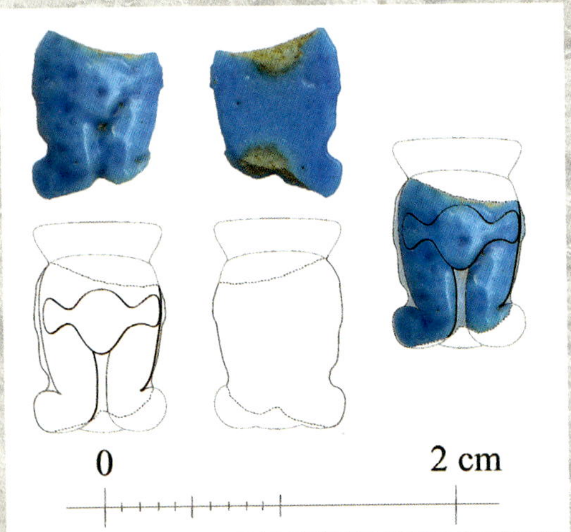

Fragments of a figurine of the goddess Hathor from around the 7th century B.C. found in the excavations of Cerro de San Vicente. Courtesy of Macarro Alcalde, C.

This archaeological site is the origin of the city of Salamanca, and the oldest vestiges found date back to the Cogotas I culture of the Late Bronze Age. However, the most noteworthy remains, and which point to a stable settlement, are from a town from the **Early Iron Age** associated with the Soto de Medinilla culture, an area which underwent constant occupation with overlapping towns from the **7th to the 4th century B.C**. Thereafter, the population increased and people began to abandon the settlement, moving to the neighbouring Teso de las Catedrales, which gave rise to the fortress of **Helmantike**.

Towards the end of the Early Middle Ages, centuries after the town had been abandoned,

thereby guaranteeing its preservation, the **monastery of San Vicente** was founded, regarded as being the first convent in Salamanca, and which was donated to the Order of Cluny by King Alfonso VII in 1143. In the early 16th century, the convent was turned into a university college and renovated for the purpose. Further work was conducted on the building over the next few hundred years, including extensions, resulting in a monumental complex of major importance. However, invading French troops turned the monastery into a fortress during their occupation, and it was destroyed after the end of the War of Independence. Monks attempted to rebuild it in the following years, but it was totally abandoned and fell into ruin pursuant to the Spanish Confiscation. The pre-Roman settlement was discovered in 1949 when construction work began on the Pontifical University Nuestra Señora de Guadalupe halls of residence, which were demolished in the early 21st century as part of the City Council´s initiative to restore the archaeological site, and which has built an interpretation centre designed to provide a better understanding of this locale.

The town was located in a flat area on a rocky escarpment of significant strategic value due to it being protected by the Tormes river and the Los Milagros stream, which has since dried up, and which is likely to have served as an important source of water. The weakest part of the defences was reinforced with a 90-metre long wall, although remains of buildings have also been found beyond this wall due to demographic pressure.

The most noteworthy of the buildings excavated so far are the houses, one of which is considerably bigger than the others, suggesting it was inhabited by one of the town´s elite families, in addition to outbuildings such as barns and ovens. These houses were made largely of clay, although stone was also used. Most of them are round-shaped, with a bench inside running around the wall that served as a resting place and an open fireplace in the middle. All these buildings were laid out around a street and it is believed that up to 250 individuals could have lived in the town.

The archaeological remains discovered include hand mills, pottery and animal bones, which give us an idea of the way in which these settlers lived, and whose economy was based on agriculture and livestock breeding. Indeed, the 2021 archaeological campaign led by archaeologists Cristina Alario, Carlos Macarro and Antonio Blanco included the discovery of an amulet of the goddess Hathor and other objects of Egyptian and Phoenician origin, thereby providing us with an insight into trade relations with eastern Mediterranean civilisations through the ancient Vía de la Plata connecting the north and south of the Iberian Peninsula.

SALAMANCA AND ST. CEBRIÁN CAVE 71

Of the former Romanesque church of **Saint Cebrián** all that is left are some remains of its crypt, taking advantage of the uneven ground. The importance of this temple from the **12th century** is rooted in being, according to tradition, the place where, during the Middle Ages, a **school of Necromancy** was located in which the teacher was the devil himself. According to tradition, in the crypt or cave the devil would give his classes in the dark arts to seven students, with the most applied of them remaining at the service of the devil. That's where the story of the writer **Enrique de Villena** arises, the man without a shadow, who was the most applied student in his final year; legend has it that not being able to abandon the cave, taking advantage of the devil having gone out, the young man hid in a pot and when the devil came back, realizing that the young man wasn't there, he started looking for him leaving the door ajar which Enrique took advantage of to leave, but at the last moment the devil found him but could catch his shadow, hence the nickname of the man without a shadow.

The fame of the place was so important that **Miguel de Cervantes** himself dedicated a sketch to him entitled *The Salamanca Cave*.

IRON AGE WALL 71

SEDES ANTIQVA CASTRORVM

In the 4th century B.C., the settlement founded in Cerro de San Vicente was moved to the next layer of the Cathedrals, probably due to space-related issues, giving rise to a walled Vaccei fortress from the **Middle Iron Age** known as **Helmantike**. In the summer of 220 B.C., the Carthaginian general **Hannibal** attacked and captured the indigenous city, one of Rome´s allies, during the Second Punic Wars fought between the Carthaginians and the Romans; this siege was documented by Polybius, Titus Livius and Plutarch, among others.

Remains of this city have been discovered under the historic old town of Salamanca, the most noteworthy finding being part of the wall, which is open to visitors, on Cuesta de Carvajal and which dates back to the final third of the 4th century B.C. There are also vestiges of the same wall in a courtyard on Calle del Silencio, in the courtyard of the School of Geography and History and in the cloister of La Clerecía, where remains of the moat were found, as well as in a plot of land on Rúa, where a tower was

Cave of Salamanca, wall and Enrique de Villena's tower.

SECTION OF THE WALL AT CUESTA DE CARVAJAL

- *Wall from the 4th century B.C.* (yellow)
- *Medieval wall* (blue)
- **F** *Filled-in ditch from the 4th century B.C.*
- **V** *Visible part on Cuesta de Carvajal and on this page.*

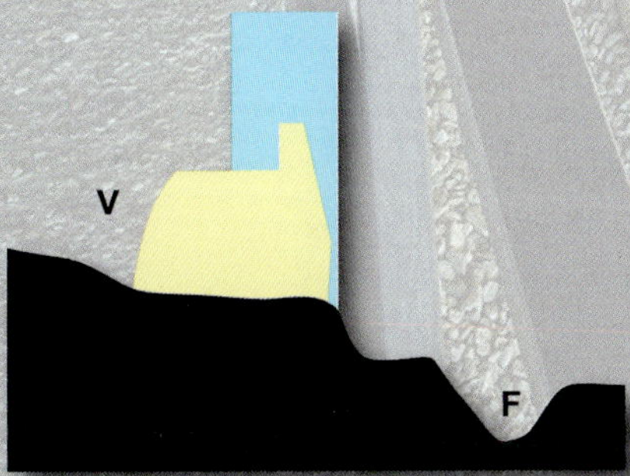

discovered, which could be part of an entrance to the fortress, and part of the wall. Studying all these remains gives us an idea of the original layout of the pre-Roman wall and therefore of the ancient city. This wall was made of masonry and featured a deep ditch dug into the rock, of which there are hardly any remains due to the pre-Roman wall having been modified by the medieval wall running parallel to it. The most striking feature of the wall is its thickness, which measures 7 metres in this area, although it was 3 to 4 metres thick in other areas.

A fence dating back to the 4th century B.C. at cuesta de Carvajal.

SALAMANCA
AB PONTE SALMANTINI
Arroyo del Zurguén
Side of the road where the grass still tends to grow less than in the ditches of a greener appearance.
Foundations of the Roman road consisting of aggregate and pebbles, from which the aggregate of the surface layer has been removed to make them visible.
AB AVGVSTAM AEMERITA

VIA DE LA PLATA 1

VIA XXIV · ITER EMERITA AB ASTVRICA

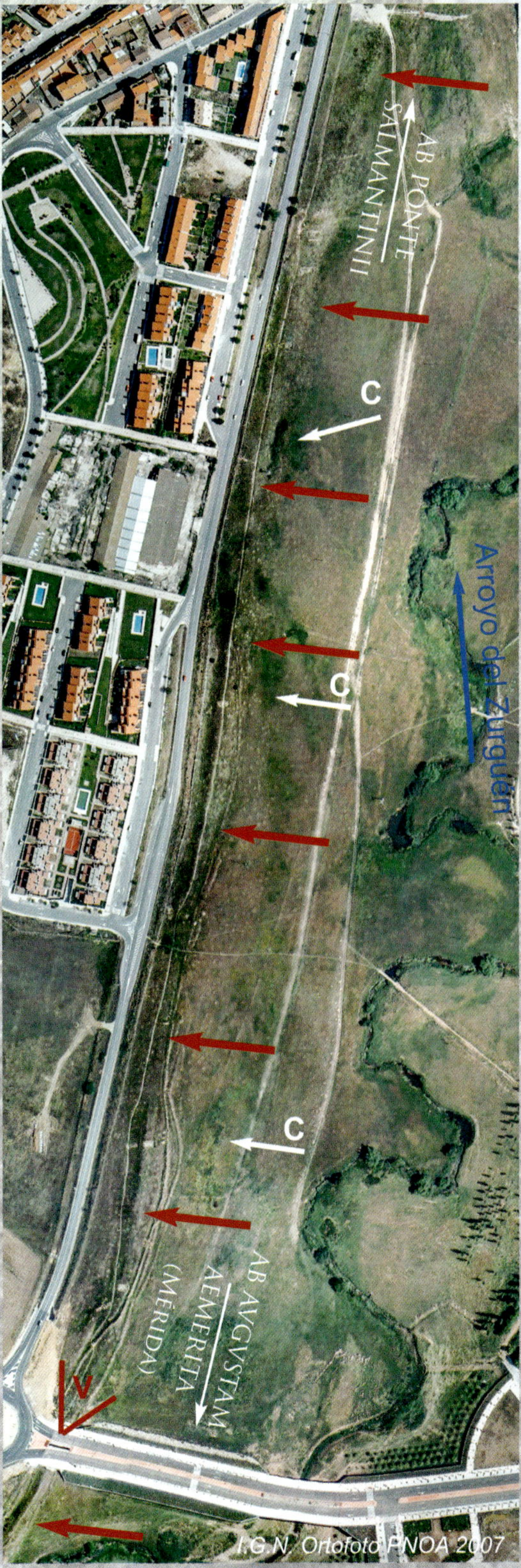

The route known as Via or Ruta de la Plata is a 20th century creation of a tourist and cultural nature that connects the north and south of western Spain from Gijon to Seville along the A-66 motorway. The current route runs along trading and communication paths used in the region in ancient times. In pre-Roman times there was a route that connected Helmantike (Salamanca) with the main Tartessian and Phoenician settlements in the south along which minerals such as tin and other products imported by these towns were traded. Later, in Roman times, there was a road that linked two of the main Roman cities in the Peninsula: Augusta Emerita (Merida) and Asturica Augusta (Astorga), the route of which, included in the Antonino Itinerary, does not exactly coincide with the current parallel roads, and many sections are 15 to 20 kilometres away. The case of Zamora, a landmark on the current route, is paradigmatic, as it didn´t exist in Roman times due to the fact the Roman road led to Ocellum Durii, Villalazan.

Its name arises from phonetic evolution, although there are various hypotheses in this regard. One of them interprets the name *Plata* as coming from the Arabic *al-Balat*, meaning paved road, although more recently it is believed to have come from the term *via delapidata*, which would translate to a path marked with milestones, which is far more likely as Roman roads were only paved in urban areas. Part of the road and the Roman bridge are still visible in the outskirts of Salamanca.

Layout of the backbone of the ROUTE

C Roman quarries at the foot of the route for the extraction of materials, now wetlands.

V Direction of view of the photograph

For more information on this and other routes in Castile and Leon, you can refer to the excellent free scientific work of Isaac Moreno Gallo.

Modern 17th century factory on the bridge, made of granite and sandstone.

ROMAN BRIDGE 1

This formed part of the Vía Palata (Silver Road), the **24th Road on the Antonine Route** which joined Mérida and Astorga. Although its origin is Roman, **1st century AD**, its complete original construction work has not been retained as it has undergone much remodelling work and reconstructions as a consequence of the overflows of the river, the most important is that known St.Polycarp in 1626; hence, the **15 archest** nearest the city are original which have been carried out using padded granite ashlars. In 1626 the river dragged the other 10 arches of the bridge which is why it had to be reworked, with the old and the modern factory being separated by the central turret. The bridge was battlemented until the 19th century.

The Roman construction system, applied later with some modifications to the rebuilt area, is that of a small pillar with starling upstream and the springers of arches; the space that remained free in the pillars under the arches served to support the formwork which was to secure the arch until the latter was finished. In the majority of ashlars you can still see the holes of the ferrei forciceps, or clamps, which were placed in the latter and thanks to the very weight of the ashlar remained strongly secured and could lift it up.

Holes in which the ferrei forciceps were inserted

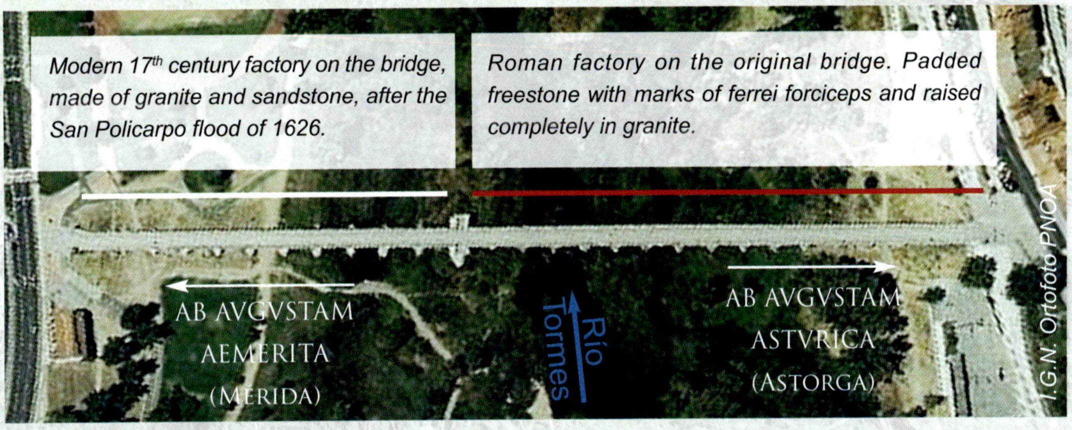

Modern 17th century factory on the bridge, made of granite and sandstone, after the San Policarpo flood of 1626.

Roman factory on the original bridge. Padded freestone with marks of ferrei forciceps and raised completely in granite.

This part of Salamanca, together with the valley of La Palma and the Cerro de San Vicente were completely destroyed in the war against the French invaders, and were only fully restored in the late 20th century. The loss of the schools of Oviedo and Cuenca, in addition to the convents of San Agustin and San Vicente, was particularly regrettable.

New Cathedral
University of Salamanca
Pontifical University
Museum of Salamanca
Ruins of San Pedro
Cloister
RUINS OF THE OLD CONVENT OF ST. AGUSTÍN

THE CATHEDRALS OF SALAMANCA

OLD CATHEDRAL OF ST.MARY'S OF THE SEAT

NEW CATHEDRAL OF THE ASSUMPTION OF THE VIRGIN

The Cathedrals from the bank of the Tormes

GROUND PLAN AND CONSTRUCTION STAGES OF THE CATHEDRAL

OLD CATHEDRAL

- Romanesque, 12th to 13th c.
- Gothic, 14th century
- Gothic, 15th century
- 16th century
- Neoclassicism, García de Quiñones, 18th c

NEW CATHEDRAL

- Juan Gil de Hontañón, fin 1520
- Juan de Álava, fin 1520
- Rodrigo Gil de Hontañón, 1538
- Juan Setién Güemes, 1668
- Pantaleón Pontón Setién, 1703

ADDITIONS AND REMODELLINGS

- Manuel de Larra Churriguera, mid-18th century
- Baltasar Devreton, 1755
- Juan Setién Güemes,

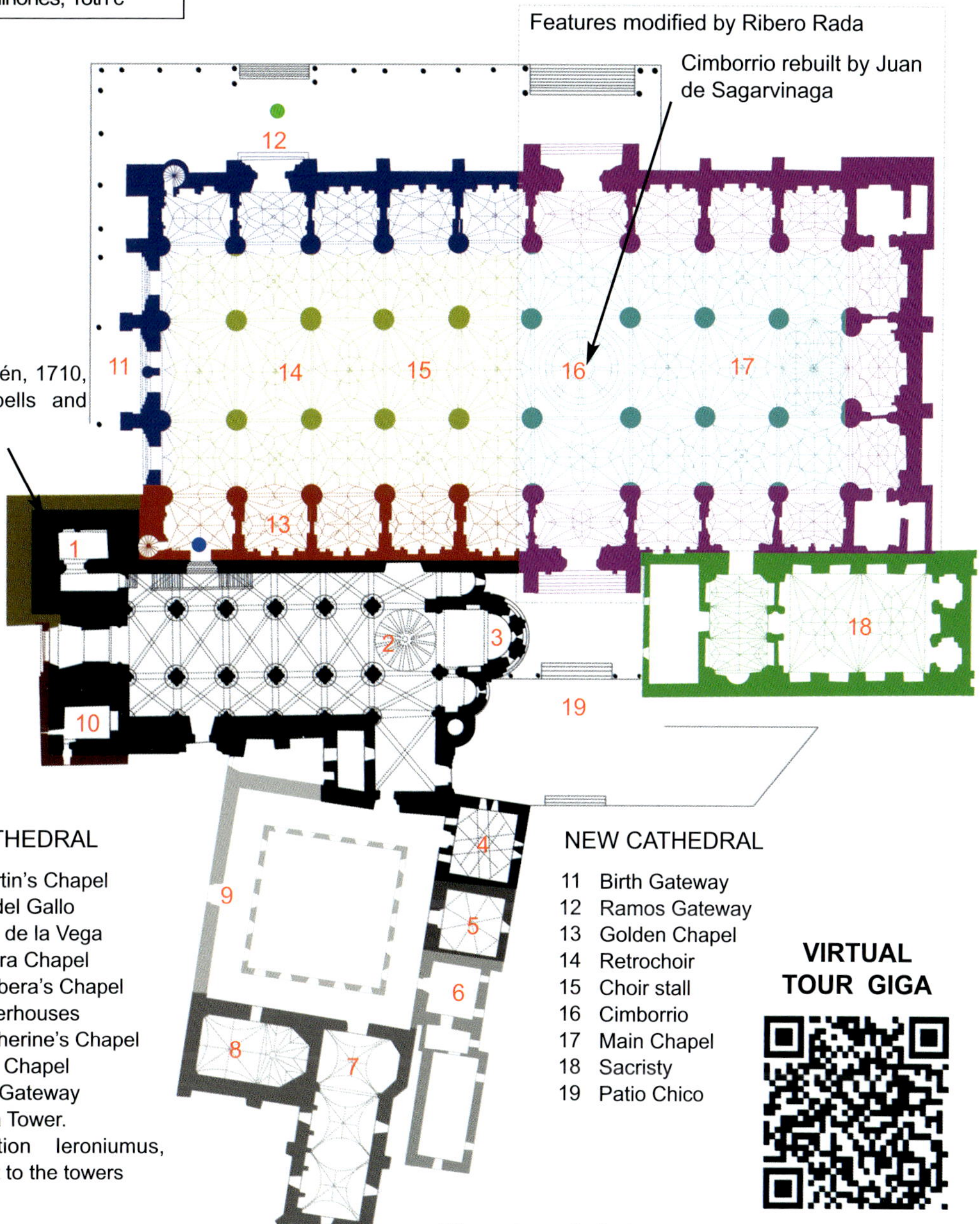

OLD CATHEDRAL

1 St.Martin's Chapel
2 Torre del Gallo
3 Virgen de la Vega
4 Talavera Chapel
5 St.Barbera's Chapel
6 Chapterhouses
7 St.Catherine's Chapel
8 Anaya Chapel
9 Carts' Gateway
10 Mocha Tower. Exhibition Ieroniumus, ascent to the towers

NEW CATHEDRAL

11 Birth Gateway
12 Ramos Gateway
13 Golden Chapel
14 Retrochoir
15 Choir stall
16 Cimborrio
17 Main Chapel
18 Sacristy
19 Patio Chico

VIRTUAL TOUR GIGA

- Entrance to the New Cathedral
- Entrance to the Old Cathedral

The city of Salamanca is lucky to have two cathedrals which together incorporate all the architectonic styles from Romanesque to Neoclassicism.

OLD CATHEDRAL OF ST. MARY'S OF THE SEAT

After the official resettlement in 1102 the diocese was remodelled and the need was raised to build a seat to which end the plot of the former Visigoth main church was chosen. The **Old Cathedral** building started to be planned between 1240 and 1250, presenting three basic stages: the first corresponding to the years 1170 and 1175; the second would extend from 1175 to 1200-1210; and the last occupied the years from 1200 to 1230.

The architectonic system deployed is still Romanesque, but the language is now late Romanesque with pointed arches with considerable arrow, dual pointed arches and the ribbing barrel vault with pointed arches.

The temple began to be built via the head and simultaneously the base. It has a Latin cross ground plan with three naves which end in semicircular apses. It retains the enclosed arches of the presbytery, typical of full Romanesque style; but later in the presbytery stretch innovations are introduced like the pointed barrel vaults and in the rest of the ribbing.

The central nave has a pointed ribbing vault, but the starting point varies; this vault rests on corbels which seek to conceal the joint of the ribs with the wall. The supports are pilasters with back-to-back columns or semicolumns on which the transverse arches rest and there are also other crab columns. This cathedral was designed to be covered with a barrel vault in the central nave and arris on the sides, but its alteration meant the support also had to be changed.

In the **16th century** it was greatly altered because of the construction of the New Cathedral, it being considered that the Romanesque had become too small for the importance possessed by the city; it was initially thought about totally knocking it down, but finally the new construction was built back-to-back in the northern area, with part of the nave and the apse disappearing, still perceptible both to the interior and exterior from the Patio Chico.

The **western gateway** disappeared in **1679** when it was replaced by the dull façade erected by **Juan de Setién Güemes**. The towers which flanked it also underwent modifications; the southern one was not finished and the northern one was lined after the Lisbon earthquake in 1755 when its structure was greatly damaged, it had already been modified previously to use it in the New Cathedral, adding the last bodies in the early 18th century which were carried out by Pantaleón Setién who tried to integrate the Baroque architecture with the preceding Gothic and Romanesque ones.

Capitals of the apse windows with fantastic beings

LA TORRE DEL GALLO

At the transept we can find the well-known **Cockerel Tower**, a turret inspired by that to be seen at **Zamora Cathedral**, the seed for the new model of French roots. At the beginning it was thought about not covering this space with a turret of the size of the current one, but rather with a ribbing roof.

The profile changes with regard to Zamora, bulbous in shape, and this more stylised than the former; it has two storeys of windows in the drum and is strengthened by four turrets. Its conic structure is adorned by scales of stone and stonework which separates the rowlocks from the scale covering, but the latter do not correspond to the interior.

The interior bears a gored dome on pendentives which, contrary to the one at Zamora, is separate from the exterior. The gores are separated by ribs which rest on the columns which serve as separation between the windows.

Over time it was dismantled piece by piece owing to its poor condition and this may be the reason why its arrow profile was boosted to lend it a more Gothic appearance.

Sección longitudinal del cimborrio por el centro. Dibujo de Stüler, Enrique. R.A.B.A.S.F.

EVOLUTION OF THE LATE ROMANESQUE CIMBORRIOS IN THE AREA

VIRTUAL TOUR GIGA

Cimborrio at Zamora Cathedral, 12th century

Cimborrio at Salamanca Cathedral, 12th-13th centuries

Cimborrio at Toro Collegiate Church, Zamora, early 13th century

Cimborrio at Plasencia Cathedral, 13th century

Cimborrio at La Lugareja church, Arévalo, Ávila, mudéjar 13th century

THE INTERIOR

Inside the temple its rich **sculptoric decoration** stands out, located in the capitals and corbels. The ribs of the Gothic cross vaults rest on **corbels** with representations of monstrous animals. On the **pendentives** of the base there are angels on masks with diabolic connotations. Also worthy of mention are the capitals of the naves with high stylised phytomorphous decoration, highlighting some historiated ones like those representing Samson fighting the lion or that of Adam and Eve eating the forbidden fruit.

At the **transept** we can find several **Gothic tombs** decorated in excellent reliefs, sculptures and paintings. They belong to different ecclesiastical posts or benefactors of the Cathedral.

St.Martin's chapel, also called the "Oil" chapel, is situated at the foot of the temple under the tower body. The walls of the top guide are decorated by **Gothic murals** carried out by **Antonio Sánchez de Segovia** in **1262**; they form an excellent collection which illustrates the pictorial work of the Early Gothic. In the centre a niche opens out, now empty, which would have a sculptural image of the Virgin around which there are portrayals of angels and prophets under Gothic architectures, accompanied by San Joaquin and St.Anne. These paintings are brightly coloured with very dynamic postures and garments. On the adjoining wall there are other lower quality, less colourful paintings which, like a tapestry, cover it with the theme of the Last Judgement and the Etimasia or Expectation.

The **main chapel** lies presided over by a colourful **retable** attributed to **Dello Delli**, a Florentine painter from the first half of the **15th century**; it comprises 53 painted boards in the temple, organized into level aisles and five bodies which recount the story of Jesus. These boards are characterised by presenting a style halfway between the Gothic and the Renaissance; though the postures and treatment of the figures, as well as the colouring recount the best of the Gothic, the architectures serving as the stage for the various landscapes already depict a relationship with the Renaissance.

In the centre of the main retable there lies the **Virgen de la Vega**, the patron saint of the city, which was brought from the Monastery of St.Mary's of the Fertile plain. It is a late Romanesque sculpture from the early **13th century** with a wooden core and lined in golden copper adorned by cabochons of gems and enamels from Limoges.

On the main retable, in the oven arch, there is a depiction of the **Last Judgement** with an al fresco painting carried out by **Nicolás Florentino** in the mid-**15th century**; this painting, contrary to the retable, depicts a full adaptation to the new Italian Renaissance models.

THE CLOISTER

The **cloister** is located to the south of the Cathedral; around it the Chapterhouses and various chapels open out. The original Romanesque cloister was remodelled in the late **18th century** by **Jerónimo García de Quiñones**. It seems to have shown arches of coupled columns, similar to those kept at the **Monastery of Santa María de la Vega** whose origin is unknown and they may belong to this monastery or be the remains of the cloister of the Old Cathedral, brought here after the aforementioned remodelling; the unknown aspect has yet to be cleared up, what we do know is that stylistic characteristics linked with the Cathedral can be observed.

The first chapel we can find is the **Talavera chapel**, also called the Saviour Chapel or the Mozarab rite chapel; it is the oldest of them all and it stands out for its starry vault in Romanesque-mudejar style. It is here where **Rodrigo Arias Maldonado** is buried, born in Talavera, hence the chapel's nickname. This character donated goods so that the Mozarabe rite could be celebrated there. Hanging there is the **Banner of the Comuneros** by Francisco Maldonado, the grandson of the former and one of the leaders of the revolt against Carlos I.

Then we can find the **St.Barbara's chapel** where the **University degree exams** took place; it was here too where the dean was elected and proclaimed, all due to the fact that the Cathedral was the origin of the University[52]. It is from this time that we get the expression ***"estar en capilla"*** (Spanish expression meaning awaiting execution whose literal meaning is "to be at chapel") as those aspiring to be a doctor or to gain any other title had to spend the night before the exam looking at their books in this chapel sitting on a seat with their feet supported on those of the tomb of Bishop Lucero. If he passed the exam, he was taken via the main gateway of the Cathedral and a party was

held at which a bull would be fought, but only in the case of those gaining PhDs and with their blood and almagro, the anagram of Víctor (victoria) would be written on the wall with the name of the new doctor. If, on the other hand, the aspirant failed to pass the test, he would leave by the Carts' Gateway.

At the **Chapterhouses** several boards were kept carried out by the great Hispanic-Flemish painter **Fernando Gallego** who was active during the 15th century in the area of Zamora and Salamanca, being the maximum representative of this style in Castile. They include that known as the **Virgin of the Rose**.

Another of the chapels opening out onto the cloister is that of **St.Katherine** whose interior conserves numerous objects related with both Cathedrals.

Finally, there is the **chapel of St. Batholemew's or Anaya**, founded by Diego de Anaya, the archbishop of Seville and founder of the first Spain University, the one known as St.Barthelomew's or Anaya[39], situated opposite the New Cathedral. In the centre of the chapel there lies his tomb, made of alabaster and surrounded by a grating from the 16th century. The image of the recumbent is attired in garments characteristic of his dignity; accompanying him there is on the header an angel and a prophet, the symbol of truth, and at his feet three animals: a hare, a lion and a dog which symbolise vigilance, energy and faithfulness. Various scenes adorn the box of the tomb; at the header there is a Calvary, on one side a portrayal of Christ with the cross, the Gospels and the apostles and on the other side the Virgin with twelve saints. At her feet there are two angels holding up the coat-of-arms of the Anaya family. On the walls of the chapel there are various sepulchral tombs representing members of the Anaya family.

St.Barbara's chapel.

NEW CATHEDRAL OF THE ASSUMPTION OF THE VIRGIN

HISTORY OF THE CONSTRUCTION

In 1497 it was decided to build a new Cathedral on a larger scale; in 1509 it took shape and its initial features were provided in 1510 by Enrique Egas and Alonso Rodríguez, with the works having been started in **1513** with **Juan Gil de Hontañón** as the master builder. It was decided not to pull down the Old Cathedral so that worshipping could occur there during the construction of the new building. It was carried out on a piecework basis, with several master stonemasons working at the same time; hence, in 1520 Juan Gil de Hontañón and Juan de Álava worked at the same time on the chapels on the buttresses of the bases; the chapels by Juan Gil are more Gothic with pointed arches and vaults endowed with a pronounced curve, whilst Juan de Álava carried out flatter vaults with Renaissance frames.

Upon the death of Juan Gil de Hontañón, **Juan Gil el Mozo** took charge of the works, being replaced on his demise by **Juan de Álava** until 1537 who added the features of the main doors and the main façade has been attributed to him too.

Rodrigo Gil de Hontañón managed the works between 1538 and 1567, focusing his intervention on the vaults and the spans from the naves to the transept.

Its construction continued until the 18th century, but it still retained its Gothic roots. Here, in Salamanca, as with the Cathedral in Segovia, where the Salamanca solutions and model took pride of place, we encounter a building which, although it is erected following the late Gothic features, is carried out in a century which corresponds to the Renaissance, the 16th century which is characterised in Spain

The New Cathedral from the towers of the Clergy

as it does not have, as in other European countries, purely Renaissance architecture as scarcely any time went by between the last throes of the Gothic with Renaissance solutions and the new classicist style, already framed within the early Baroque; in this way a purely Spanish style was created.

It began to be erected via the gable of the bases; the initial Project consisted of a large room ground plan with three practicable naves and two sides dedicated to chapels, as well as a polygonal ambulatory at the header. When **Juan de Ribero Rada** took over the works during the second construction stage, the features were changed in the late 16th century, now incorporating classicism, in vogue at that time and the ambulatory became rectangular; little was done during this period owing to the scarcity of economic resources. Until the second half of the 17th century the Works were not resumed and the master builder was **Juan de Setién Güemes**; at this time the gateways of the transept were carried out and the chapels of the ambulatory. He was followed in the post by **Pantaleón del Pontón Setién** who closed the vaults of the transept and header, as well as building the upper bodies on the former Romanesque tower. The carrying out of the turret remained pending which was undertaken by **Joaquín de Churriguera**, though after the Lisbon earthquake it had to be dismantled and rebuilt, work which was undertaken by **Juan de Sagarvinaga**.

BIRTH GATEWAY

On the **exterior** worthy of special mention is the main façade located at the base of the temple; this has five engrailed arches which correspond to each of the naves. The three central arches boast various sculptural and architectonic decoration with the succession of different kinds of arches, but the maximum sculptural expression is to be found in the central one where the architectonic elements, the figures and the reliefs take up the whole space in an expression of horror vacui like a tapestry. The structure of this gateway known by the name of **Birth Gateway** is purely Hispano-Flemish; it is believed to have been designed by **Juan Gil de Hontañón** and it was worked on by various masters including **Gil de Ronza**. However, this gateway remained incomplete except for its

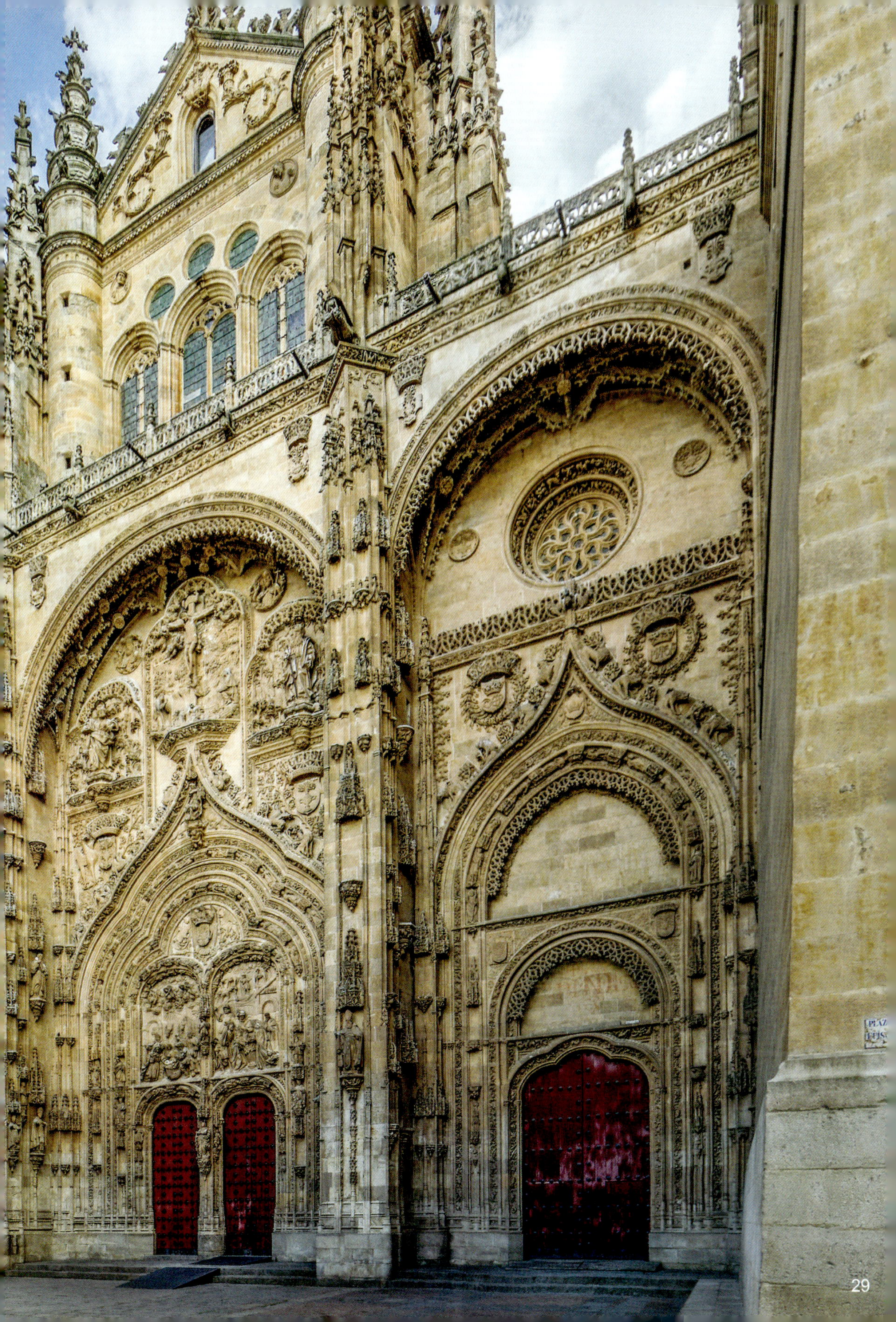

structure and the Calvary relief, the rest was carried out in the mid-17th century; this age involved the great reliefs of the Birth or the Adoration of the shepherds and the Epiphany which show a way of working which is very distinct from that of the late Gothic; the figures are very expressive and naturalistic in their gestures and postures and the clothes are folded, lending dynamism and the theatrality characteristic of the Baroque to the whole. Also part of this age and by the same author, **Juan Rodríguez**, a disciple of Gregorio Fernández, are the figures of St.Peter and St.Paul which flank the Calvary and the Immaculate Virgin located at the mullion.

Immaculate Virgin

GIGA PHOTO FAÇADE

The Birth and the Epiphany.

Gateway tympanum

RAMOS GATEWAY

On the northern side the **Ramos Gateway** opens out whose relief of the Triumphal Entry into Jerusalem was also carried out by **Juan Rodríguez**, repeating the characteristics of the reliefs of the Birth Gateway. This author is also behind the sculptures of the gospels located on the side which stand out for their naturalism and the generous nature of their garments. As a curiosity it is worth adding that the decorative motifs of this Gateway include some portrayals such as that of an **astronaut**, a lynx, a bull, a dragon with an ice-cream, a river crab, a stork or a hare; they are the fruit of restorations undertaken in the last few years of the 20th century which represent the tradition of incorporating new elements in each of the restorations which the building has undergone.

The astronaut

The dragon with an ice-cream

The river crab

THE INTERIOR

Internally, the New Cathedral stands out for its amplitude and grandiousness, compared on many occasions with Seville Cathedral.

As we come in we come across the fabulous **retrochoir** designed by the **Churriguera** brothers, a living example of their way of working with a lot of decoration that covers the whole space. It is designed as a true retable with three aisles finished off by a shell with the relief of God the Father surrounded by heads of angels and close-knit decoration. On the side aisles there are two sculptures representing **St.Anne** and a **St.John the Baptist**, works by **Juan de Juni** from the Old Cathedral.

Another by the **Churriguera** is the **choir stall**, whose walnut wood stalls were designed by Joaquín. It consists of 104 ceremonial chairs; the ashlar reliefs were carried out by **José de Larra** and **Alejandro Carnicero**, and their figures are slim, elegant and very moving.

Within this space special mention should be made of the **two organs**; one was carried out in the 16th century and the other in the 18th century by the famous royal organ-maker **Pedro de Echevarría**.

Of the seventeen chapels surrounding the Cathedral worthy of special mention are the so-called **Golden or All the Saints Chapel**, built by Juan de Álava. Hundreds of saints bedeck its walls and the Golden polychrome of the latter and of the architectonic elements of the tombs are what lend them this name.

The other chapels accommodate sculptures carried out by great masters of all ages such as the **Piety** by **Luis Salvador Carmona** at the Sacramento chapel, the **Virgin of Carmen** by **Antonio de Paz** at the Christ of the Battles chapel, or the **Virgin of Solitude** at the chapel of the same name carried out by **Mariano Benlliure**.

Finally, it is worth mentioning the most beautiful element of this Cathedral: its **dome**, originally carried out by **Joaquín de Churriguera**; also by this author there only remains the scallops and the lower drum as it had to be dismantled as a result of the **Lisbon earthquake** which seriously damaged it. Its reconstruction was carried out by **Juan de Sagarvinaga** who finished the dome. It looks splendid with its adornment of reliefs and it lends greater luminosity to the space.

EL MARIQUELO

This character, who, on the eve of the festival of All Saints Day, goes up to the tower, receives his name from the surname of the family of bell ringers who went up for the first time after the Lisbon earthquake to check the damage suffered by the tower and to give thanks to God for avoiding any greater deaths to the Cathedral.

Even today, though it is not the same family, other people have taken up the testimony and the same ascent is carried out on said date; congregating the wealth of Salamanca dwellers and any curious onlookers of the tower.

IERONIMUS

Ascending the tower

This is an exhibition and visit to the Cathedral tower and from here all the platforms interiors can be observed boasting wonderful views from the naves, buttresses or the Cockerel Tower. In the interior, plans and a documentary by the construction process, ruin and rebuilding of the bell tower.

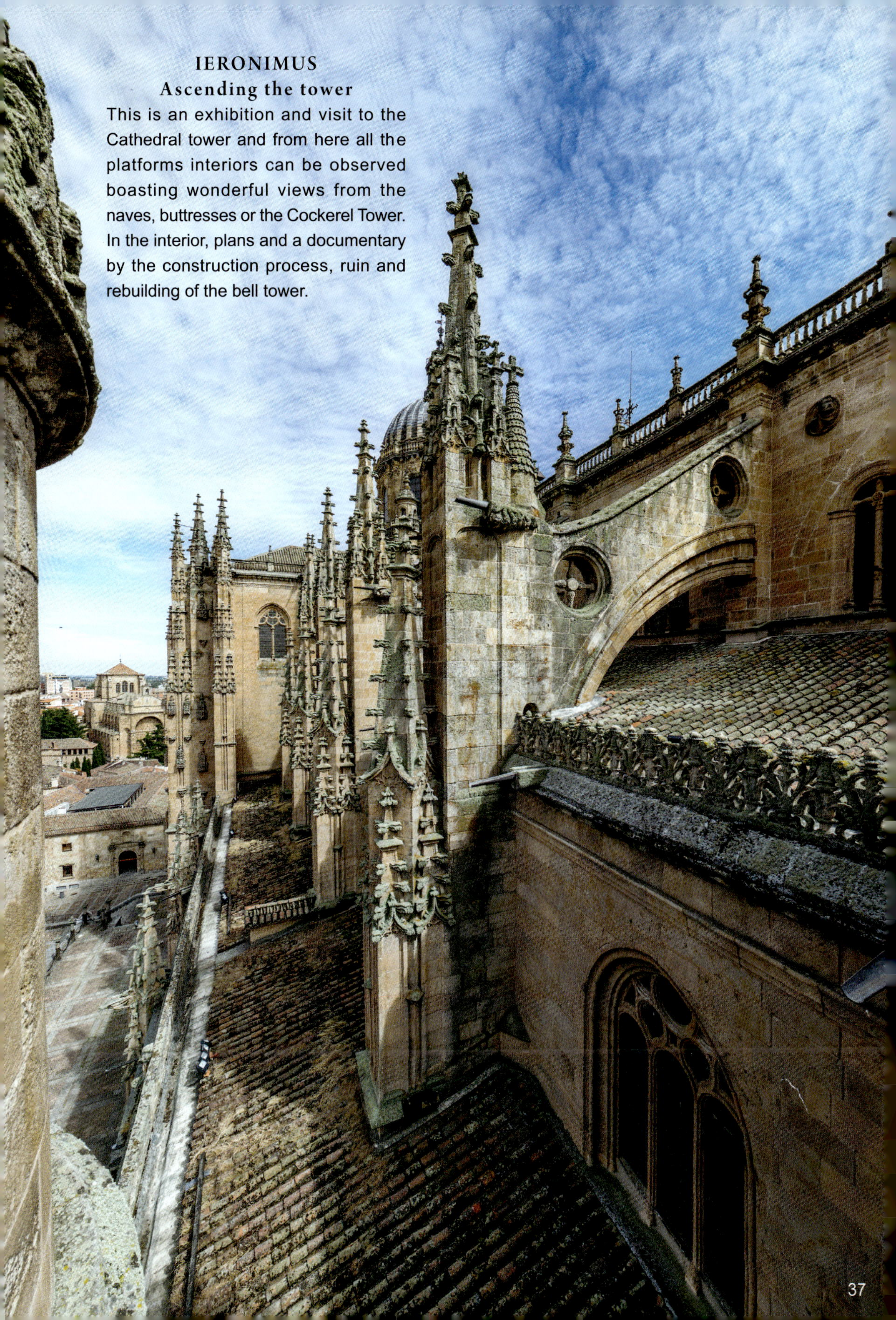

PLAZA DE ANAYA AND UNIVERSITY COLLEGE OF ANAYA

CHURCH OF ST. SEBASTIAN[69]

The former chapel of the St.Bartholomew's School was built by **Alberto de Churriguera** in **1731**. It has two intricately decorated gateways of Carmelite inspiration; the main one comprises three bodies, with the central one occupied by a sculpture of **St.Sebastian**. The side gate follows the same scheme as the main one, but the statue represents St.John of Sahagún, studied at St.Bartholomew's, under blinds drawn by angels. The sculptures of both Saints were carried out by **José de Larra**.

In the interior there is a Latin cross whose nave is covered by an intricately decorated arch of lunettes profusely decorated. On the transept a large octagonal dome of Madrid roots opens out.

The original main altar by **Simón Gavilán Tomé** was dismantled and at present it is divided between St.Martin's Church[122] and the New Cathedral[26].

UNIVERSITY COLLEGE OF ST. BARTHOLOMEW'S OR ANAYA[68]

This is the oldest University College in Spain; it was founded by Diego de Anaya y Maldonado, the archbishop of Seville in 1401 on the plot taken up by the church of St.Bartholomew's.

The original building began to be built in the first quarter of the 15th century; in the mid-16th century remodelling works were carried out under the command of **Alonso de Covarrubias**; said building was endowed with its own chapel and library. This was all demolished to erect the imposing construction we can see today, built in two stages; the first was undertaken between the late 17th century and the early 18th century, involving the northern and eastern wings. The second part was carried out in accordance with a design by **José de Hermosilla** with the work being managed by **Juan de Sagarvinaga**. This is line with the new Neoclassical taste; the entrance raised by a flight of stairs is adorned by four large columns which support a majestic front in the style of the old classical temples. The sole element breaking with this new look is the huge coat-of-arms

crowning the building whose volume and movement are reminiscent of the Baroque.

On the **courtyard** there is the same scheme endowed with large proportions and classical taste in its dual height and huge columns on the lower floor; it is influenced by the Palace of Carlos V at La Alhambra. Also worthy of special mention is the majestic **imperial stairway** which reminds us of Baroque theatricality and inspired by the stairway of the Royal Palace of Madrid. This seems to be presided over by a bust of Miguel de Unamuno carried out by the sculpture from Palencia **Victorio Macho**.

Its most renowned students are St.John of Sahagún and Alfonso de Madrigal.

ANAYA GUEST HOUSE [67]

The **Anaya Guest House** was built as a residence for St.Bartholomew's school to provide accommodation to any students who had already completed their studies and were looking for work. It was built by Pantaleón del Pontón and **Joaquín de Churriguera** who took over the work after the death of the former in around **1715**.

In the exterior it pans out into two storeys with windows and balconies flanked by carved pilasters.

The interior accommodates a double-height courtyard which was left unfinished as only two of its sides were totally completed; here a strong Renaissance influence is present, not at all related with the works of Churriguera and the time it was carried out when the profuse decoration of the Baroque was at its peak.

Courtyard of the Anaya Guest House.

Courtyard of the University College of Anaya

Starway of the University College

PATIO CHICO

CALIXTO Y MELIBEA GARDEN[70]

One of the most recondite spots of Salamanca is undoubtedly this garden situated on the sections of the city wall; the place is not only evocative, designed in the style of palace gardens, but also the name which brings to mind passages of the famous work of Fernando de Rojas ***La Celestina*** which it is believed was set in Salamanca, as the author was a student of its University and said he had found an uncompleted manuscript there with the tragicomedy that he took it upon himself to complete. It is worth visiting these gardens and walking around their vegetation, enjoying the views of the fertile plain of the Tormes.

A specimen of a Canarian palm tree, extraordinary for its northern location.

LIS HOUSE.
ART NOUVEAU AND DECO MUSEUM[4]

This small urban palace is a good example of **modernist** architecture; its construction was commissioned by the entrepreneur Miguel Lis to the architect Joaquín de Vargas Aguirre in **1905**; this architect stood out for the use of cast iron in his constructions, such as, for example, the Food Market. The work carried out at this house is amazing as between the main façade and the rear there is great unevenness that the architect has been able to resolve with mastery by creating a large double-height lookout

gallery made of cast iron, decorated by beautiful stained glass windows under which a double flight of stairs which looks out onto a garden on terraces.

The original appearance of the house was modified with the addition of a glazed ceiling to the interior courtyard. For years it lay abandoned and was recovered to house the collection of art deco objects of **Manuel Ramos Andrade**, an emigrée from Salamanca, who had gradually collected pieces during his trips around Europe and he wished his collection to be deposited in Salamanca. This collection brings together pieces of various origins, highlighting the **porcelain, glass and doll** collections which enable us to gain some sort of idea of the new way of life followed in Europe in the first half of the 20th century.

House façade

Here we find the best example of civil architecture in the city after the "Conchas" (Shell) house[80]; this palace was erected in the **15th century** as the residence of the **doctor of the Catholic Monarchs**, Fernán Álvarez Abarca. This is a beautiful-looking noble house which combines elements of Hispano-Flemish art and the Renaissance, a specific characteristic of that age of transition . It stands out for its simple gateway framed by an alfiz which houses the coats-of-arms of the owners; this is displace from the central axis with regard to the rest of the façade. What stands out most about the whole entity are its windows; the central ones combine Gothic elements such as the pinnacles and the gables which crown them with fully Renaissance elements in the parapet, decorated with reliefs of dolphins facing each other. These two windows are framed by a broken spandrel which houses the coats-of-arms of the Catholic Monarchs. The windows of the ends are characteristic of the Renaissance.

Detail of a façade window

House courtyard

Room I, armour of Las Dueñas.

In the 1940's the **Provincial Museum of Fine Arts** was moved here whose collection accommodates the most diverse pieces located in the various palace rooms. Its Entrance opens out onto the Schools Courtyard, being accessed via the former garden of the house where different funeral stales and **boars** are shown, with that from **Gallegos de Argañán** standing most for its state of conservation.

Boar by Gallegos de Argañán, 3rd to 2nd centuries B.C

In the archaeology area also worthy of note is the Roman pedestal dedicated to the Emperor Caracalla whose epitaph quotes the **Ordo Salmanticensis**, in other words, the municipality of Salamanca, there being a record of its existence in the 3rd century; this was found in the School of St.Mary's of the Angels[157], reused as construction material.

Relief of Luis Salvador Carmona

The courtyard between various coats-of-arms highlights the relief of the *Apparition of the Virgin and Saint Bernard to Saint Toribio de Mogrovejo*, from the now extinct University College of Oviedo carried out by **Luis Salvador Carmona** in the 18th century.

In the interior of the building it is worth mentioning the **Mudejar coffered** and frameworked armour which covers the ceiling of one of the rooms from the Convent of Las Dueñas[150]. There are also several works by the Italian **Andrea Vaccaro**, including the figure of *St.Michael the Archangel* carried out beautifully in silver.

Finally, it is worth mentioning the deposits from the Museum of El Prado and Reina Sofía which round off the exhibition, including works by **Madrazo**, **Zuloaga** or **Carlos de Haes**.

St.Michael the Archangel A. Vaccaro

SALAMANCA UNIVERSITY[10]

OMNIUM SCIENTIARUM PRINCEPS SALMANTICA DOCET

HISTORY

In **1218** King **Alfonso IX** founded the University of Salamanca but until the 15th century the "Study" did not have its own headquarters. During this period classes were given at various outbuildings of the Cathedral Chapterhouse and the degree exams a swell as the election of the Dean were held in **St.Barbara's chapel** of the Old Cathedral cloister[8].

In **1255** under the ***ius ubique docendi*** bull universal validity was granted to the degrees granted at the University. In 1474 the first library was built, the seed of the current one. Beatriz Galindo, ***La Latina***, was its first female student and Luisa de Medrano the first woman to give a

Passage from the Patio de Escuelas Mayores to the Escuelas Menores.

class at a University. In the mid-16th century at the so-called Salamanca School, Francisco de Vitoria, inter alia, gave classes in international law or the rights of the native peoples of the recently-discovered America. Salamanca University was the model for those founded in Hispanoamerica. It was here that the Gregorian reform of the calendar by order of the Pope was studied before its application. In 1857 the **Moyano Law** granted preponderance to the University of Madrid, with only the chairs of Law, Theology and Philosophy and Letters remaining in Salamanca. In 1970 the University would gain a certain autonomy, fully regaining it in 1985 until achieving its current splendour.

The following have passed though the university's doors, inter alia: Friar Luis de León, Domingo de Soto, Antonio de Nebrija, Fernando de Rojas, Hernán Cortés, Luis de Góngora, Abraham Zacut, the 3rd Count of Olivares (the Count-Duke), Calderón de la Barca, Manuel Belgrano, Miguel de Unamuno, Jose Mª Gil Robles, Adolfo Suárez and Juan Manuel de Prada.

Suggested tour:

1 Schools and courtyard
2 Cielo de Salamanca
3 Hospital del Estudio
4 University Colleges
5 House-Museum of Miguel de Unamuno

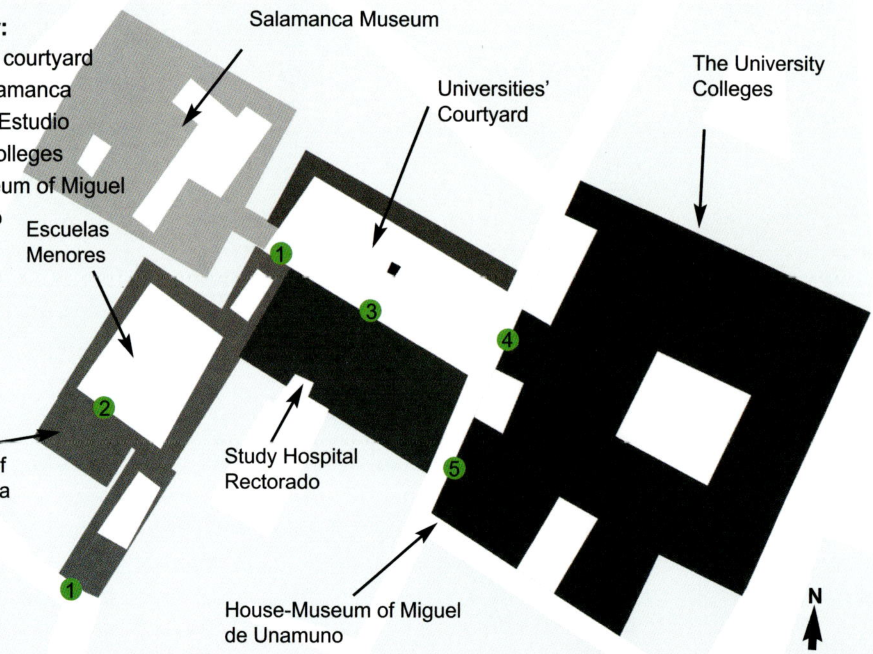

Gateway of the Schools

LAS ESCUELAS MENORES[12]

These comprise a series of outbuildings and classrooms where the studies were carried out which were needed to access the University; it is organised around a courtyard comprising **mixtilinear arches** clearly influenced by the courtyard of the Conchas House[80]. What is most worthy of mention is undoubtedly its gateway formed by two round arches on whose spandrels three medallions appear, with the central one corresponding to the image of Christ and the side ones to mythological figures. A body arises on them which, divided into three aisles, bears the emperor's coats-of-arms in the centre and on the sides of the two-headed eagles. Closing the façade a frieze runs with the papal coat-of-arms flanked by two medallions in those representing St.Peter and St.Paul. At this gate the heraldic scheme of the façade of the University College and the Study Hospital is repeated, giving iconographic unity to the whole of the University courtyard. After this façade a small courtyard opens out which provides access to the University colleges; their front is decorated by a medallion between richly adorned semicolumns on the theme of the teacher giving a class from his chair with the papal arms crowning it. Surrounding the medallion is the inscription: *SCIENTIARVM PRINCEPS SALMANTICA DOCET* already to be found on the façade of the current rectory. The rich carving of the medallions and the technical mastery make us think that the author could be Juan de Álava.

ICONOGRAPHIC PROGRAMME OF THE SCHOOLS' FAÇADE

1 Mythological character
2 Christ the Saviour
3 Mythological character
4 Two-headed eagle of Carlos I or V
5 Coat-of-arms of the empire during Carlos I
6 Two-headed eagle of Carlos I or V
7 St.Peter
8 Papal coat-of-arms
9 St. Paul

VIRTUAL TOUR GIGA

THE SKY OF SALAMANCA

In one of the rooms surrounding the courtyard what is known as the Sky of Salamanca is kept, a **mural** carried out at the temple and in oil Paint by the great painter **Fernando Gallego**, the maximum representative of Hispano-Flemish style in Castile; for a long time it was attributed to **Pedro Berruguete**, but the intervention of the latter was limited to a collaboration in which the Renaissance features of this painter can already be noticed who developed part of his artistic trajectory in Italy. This painting, carried out in around **1475**, formed part of the arch of the former library, later converted into a chapel, and it was put onto canvas in 1951 and moved to the place we are visiting today.

The authorship – or at least the participation of Fernando Gallego – is clear as the treatment and pallet correspond to his way of working such as, for example, the figure of Mercury which conforms to the human typology developed by him.

The vault bears the signs of the zodiac, the winds and some constellations and planets. Surrounding the vault there runs the following inscription taken from **Psalm 8** of the Old Testament:

VIDEBO CELOS TUOS
OPERA DIGITORUM TUORUM
LUNAM ET STELLAS
QUE TU FUNDASTI

FOR I WILL BEHOLD THY HEAVENS,
THE WORKS OF THY FINGERS,
THE MOON AND THE STARS
WHICH THOU HAST ORDAINED

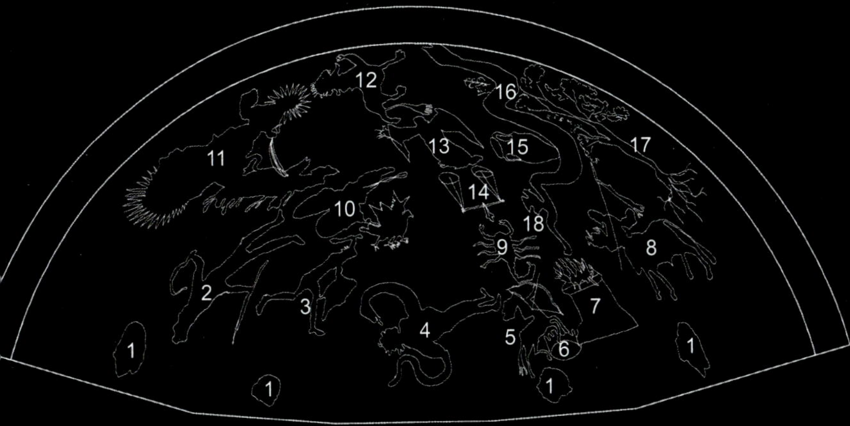

PLANISPHERE OF THE VAULT

1 Winds: Zephyr, Austro, Euro, Boreas
2 Bootes
3 Hercules
4 The Serpent
5 Sagittarius
6 Astral Crown
7 Ara
8 Centaurus
9 Scorpio
10 Mercury
11 Sun
12 Leo
13 Virgo
14 Libra
15 Rudder
16 Hydra
17 Oak
18 Crow

STUDY HOSPITAL[11]

This is the oldest building in the University College courtyard; it was built in the **15th century** as the accommodation for students on the former Jewish quarter. Of the original building all that remains on the exterior are the gateway and the windows of the chapel and, in the interior, the hallway, the stairway and the chapel.

The **gateway** is framed by a double-height alfiz richly decorated by a cardinal; on the lower part two Tuscan arches open out, separated by a column and bedecked by a semi-pointed arch whose tympanum bears an image of St.Thomas of Aquinas. On the spandrels are the reliefs of the Virgin and the archangel Gabriel, going to make up the scene of the Annunciation. On the upper part there are three coats-of-arms sheltered by arches; the central one corresponds to the kingdom of Castile and the side ones join those specific to Castile and Aragón, each supported by the eagle of St.John. These coats-of-arms symbolize the royal protection enjoyed by the institution.

Medallion of the façade

Former Hospital chapel

Gateway of the Study Hospital and the statue of Fray Luis de León

THE UNIVERSITY COLLEGES$_{10}$

The **façade** of the University Colleges is the maximum exponent of the so-called Plateresque style which boasts its own resources from the Renaissance but it still conserves Gothic reminiscences. This façade corresponds to the type called standard and its reliefs increase in volume as it gains in height so that they can be seen by the beholder under good conditions as it must be borne in mind that in the past the University College Courtyard was closed and before the gateway there was a narrow street. Nothing is known about its author or authors and there has thus been speculation about the presence of masters like **Juan de Álava** or **Enrique Egas**, but without any certainty. As regards its implementation date, neither is this known precisely but it seems highly likely that it dates back to the **1620's.**

The Catholic Monarchs

This façade is characterised by an intricate decoration endowing it with the characteristic horror vacui which was so much to the liking of the Spanish at the time. It is endowed with a highly complex iconographic programme which is hard to understand and which has resulted in the raising of different possibilities as regards its meaning. With regard to which, all the experts are in no doubt is the quality of the reliefs, probably inspired by etchings and treatises; in this regard it has been wished to attribute the iconographic design to **Fernán Pérez de Oliva**, a Cordoban humanist which is a teacher at the very University that developed the Iconographic Programme of the Fonseca School[103].

At present are two more probable theories though they derive from many sources. On the one hand, the façade is viewed as a **heraldic exaltation** through the central medallion of the first storey with the effigies of the **Catholic Monarchs**, protectors of the University as can be gleaned from the Greek inscription surrounding this medallion: *the Monarchs for the University and the latter for the Monarchs.* On the second storey there is large central coat-of-arms of Carlos I surrounded by the necklace of the Order of the Golden Fleece; flanking it is the coat-of-arms with the eagle of St.John and the coat-of-arms with the two-headed eagle of the Empire. On the last storey the papal representation as the University benefactor. The rest of the figures and medallions would refer to the virtues of the good governor by means of the representation of notable characters like Hercules, Scipio the African, Alexander the Great or Julius Caesar.

The other possibility is that it makes reference, by dint of the double door, to a counterposition between **vice and virtue** in a clear relationship with the iconographic programme of the interior stairway through which the student would be shown the right way guided by the University; hence on one side the left as you behold it, characters appear who would represent vice such as Venus, Bacchus or Priapus, and, on the other side on the right, virtuous characters like Hercules, Scipio the African or Theseus. In this regard it would be worth highlighting the role of the **frog on the skull** which would represent the fleeting nature of carnal pleasure, though at present it is lent a more mundane nature and there is the tradition that every student who finds the frog will pass in his studies.

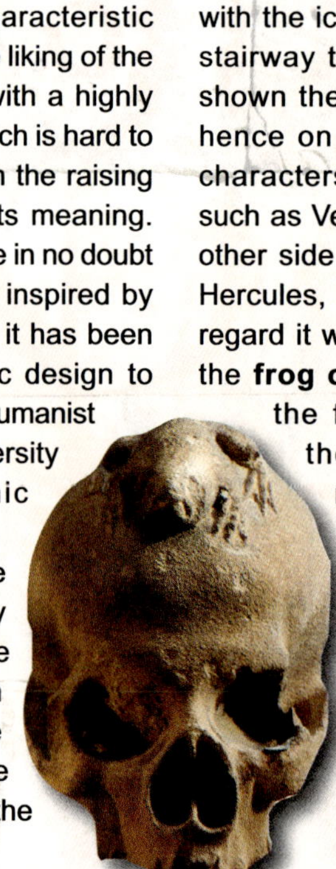

The famous frog on the Skull

Façade of the University Colleges

ICONOGRAPHIC PROGRAMME OF THE FAÇADE

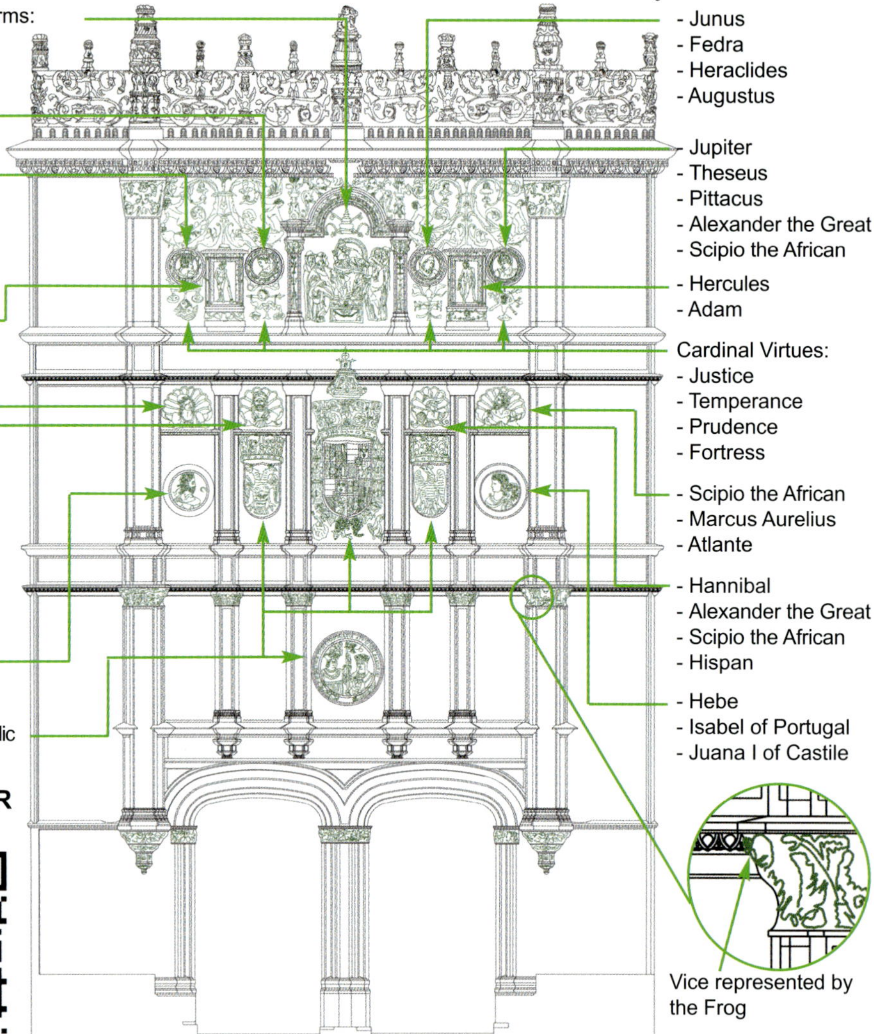

VIRTUAL TOUR GIGA

View of the façade from the original line of the street before the construction of the courtyard opposite.

PROPORTIONS OF THE FAÇADE

The Façade of the University Colleges of Salamanca University is endowed with a complex design under all the iconography which conceals it. It is wholly contained in a golden rectangle, in other words, its sides are proportional to each other through the golden number, phi or :

1,61803398874989...

irrational infinite decimal number in the periodic table. Its expression is as follows:

$$\Phi = \frac{1+\sqrt{5}}{2}$$

Present in a vast number of works of art since classic antiquity, it has been associated with the mystic and perfection; in nature itself it appears as the design of snail shells, the galaxy arms or the drawing of the inflorescences of a sunflower. It also appears in the internal decoration in a long series of rectangles drawn on the plan; it is also present in the ratio of the width between the two end columns and the lateral decoration, also marked on the drawing alongside these lines. The height of the two intermediate friezes is proportional by way of phi to the width of the columns (shown in yellow). Other parts of the decoration conform to double heights of the width such as the doors as far as the Venus and Hercules panels.

All the main iconography is included in the area of the circumference with the centre at the intersection of the diagonals which is the centre of the Crown.

If the façade is observed from the old street line, before the creation of the square opposite

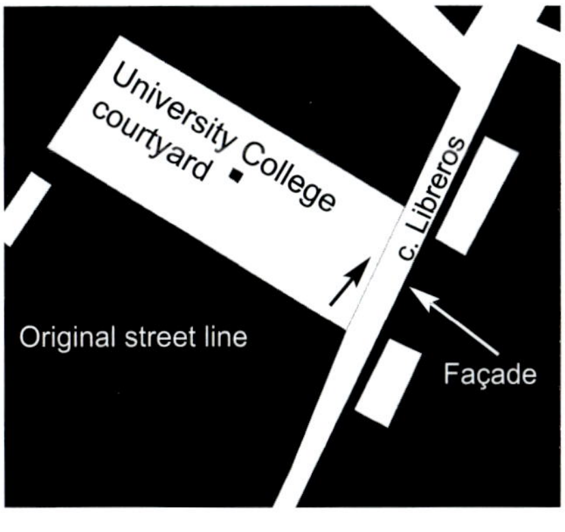

it, it can be seen how the division by friezes corresponds to an intention to visually increase the height and create a continuous progression reducing the height of the bodies on the façade until on the last storey only the flames of the largest candelabra jut out. Inside the University we will once again find the golden number.

THE INTERIOR

During the first quarter of the **15th century** construction work began on the University Colleges building which gradually expanded over the centuries, changing the features of the construction work and adapting it to the new requirements. Initially, the building had only one upper storey, having to add another to accommodate the library which required the construction of a stairway; however, of the four sides to be found on the second floor only the wing corresponding to the Library was built, the rest were added over time.

This is a square-shaped building around a courtyard where the volume of the library stands out which, furthermore, corresponds to the body of the monument's façade; the Renaissance style predominates in its construction with shades of the Gothic and the Mudejar. Top-quality architects worked on its walls such as Juan de Álava, Rodrigo Gil de Hontañón or Andrés García de Quiñones.

The entrance to the building is gained via a magnificent rectangular **hallway** covered by a tracery vault supported on Renaissance corbels. Following it, the second hallway opens out which was the original access prior to the construction of the façade and which has now been converted into a corridor providing access to the cloister. This is also rectangular in shape

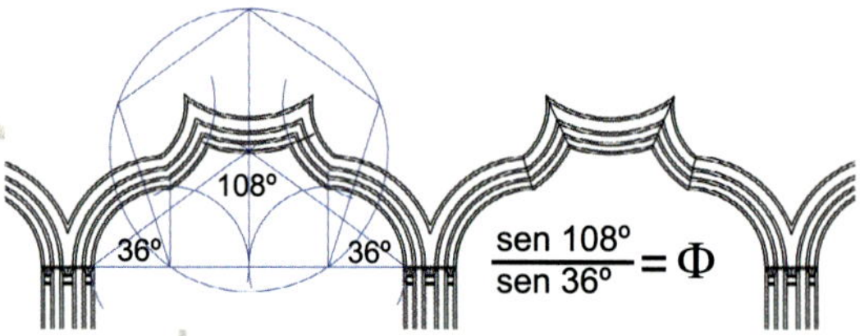

Justification of the arch layout

with two sections covered by Gothic vaults whose key parts bear a portrayal of the royal and university coat-of-arms and the rest are decorated by flowers, shells, depictions of the four Evangelists, **King David** and other prophets; the ribs of the vaults are supported on corbels with cardinal decoration.

The **courtyard** accommodates a large redwood tree brought from Canada; around it the different classrooms and the chapel are distributed on the first floor. The walls of the cloister are covered by polychromatic reinforcements under which there is a richly decorated Plateresque frieze. The upper walls, made similar to the western one or of the Enigmas, open put with arches whose features are obtained from a regular pentagon, affording an internal triangle whose sins are related through the golden number or phi which has already appeared on the façade.

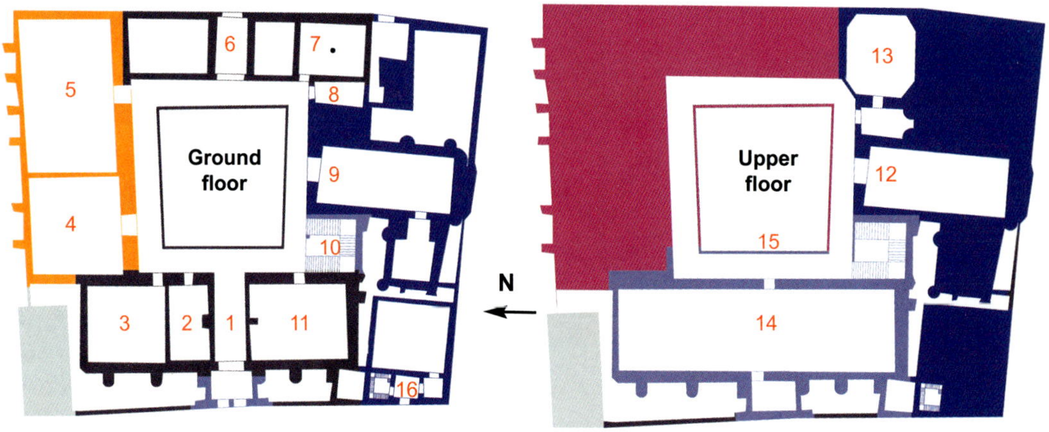

CONSTRUCTION STAGES

- First stage, mid-14th century
- Rodrigo Gil de Hontañón around 1570
- Screed, 1509
- Second half of the 18th century
- Screeds, end of the 19th century

1 Entrance hallway
2 Dorado Montero Classroom
3 Unamuno Classroom
4 Fray Luis de León Classroom
5 Paraninfo
6 Hallway, Mudejar coffered ceiling
7 Columns Classroom
8 Mercatus, exit
9 Chapel
10 Stairway
11 Salinas Classroom
12 Chapel, choir stall
13 Cloisters Hall
14 Library
15 Enigmas
16 House-Museum of Unamuno

THE CHAPEL

The room we can now see corresponds to the **third university chapel**; there had been another two, the first at a different location and the second on the same plot, but endowed with a different configuration as it had a second floor occupied by the old library which, over time, was knocked down to lend greater height to this chapel, with the paintings of the Sky of Salamanca[58] remaining as the ceiling. The current chapel shares the ground plan with the previous one, but the other elements have been lost, including the altarpiece which had boards and sculptures by Juan de Flandes and Felipe Bigarny. The new work was undertaken by **Simón Gavilán Tomé** in **1761** who carried out the design of the room and the new retable. This is an altarpiece which fits in with the aesthetics of the 18th century with a certain air that preceded neoclassicism; it is made from marbles with colours and adornments made of golden bronze. It has three aisles and two bodies to which the attic has to be added. It is a painting altarpiece which alludes to devotional aspects; **Francesco Caccianiga** was commissioned to carry out these canvases. The main body has three paintings, the lateral ones represent St.John of Sahagún in his miracle of the Yellow Well and St.Thomas of Villanueva. The central canvas depicts the **Oath of the Immaculate Conception**; the scene divides into two levels; the lower one features the teachers wearing an oath on the Bible with a notary, undertaking to defend and accept the Immaculate Conception, even before it was a dogma of Faith. The majority of characters featured in this cloister are friars. On the upper one the Immaculate Conception appears in a triumphal, apotheostic representation. In the second body there is a depiction of St. Hieronymus as the patron saint in a golden bronze medallion. Antonio González Ruiz carried out the two lateral paintings of the second body with St. Augustine and St.Thomas of Aquinas and the Calvary of the attic. Behind the central painting there is a side chapel with a sacrifice which can be seen in Easter Week or the Corpus. The painting comes down to let this space be viewed which is called a Theatre altarpiece as it recalls the theatre, a concept very much in vogue during the Baroque.

On one side of the chapel is the **cenotaph of Fray Luis de León** designed by Nicasio Seville in mid-nineteenth century.

View of the chapel from the University

CLASSROOMS

As we have indicated, the former **classrooms** or **study rooms** open out onto the cloister; on their doors, in an inscription on slate, the subject is indicated given in each: Eloquence, Canon Law, Rhetoric, Civil Law or Theology, inter alia. The classroom study of Theology, now called the **Friar Luis de León classroom**, is the one that best conserves its former appearance; the benches and desks are quite rough wooden tie beams. Around the classroom, alongside the wall, some benches have been set out from where the lessons were observed by doctors or illustrious personalities such as King Carlos I in a class given by Friar Francisco de Vitoria. Its current name derives

from the famous teacher who gave classes there and where the return of Friar Luis de León to academic life occurred after a stay in the Inquisition jail with the famous phrase: ***As we were saying yesterday***.
Another of the most representative classrooms is the study room for Canon Law, now **Paraninfo**; it is the largest one and it is presided over by the banner of prince Juan, the son of the Catholic Monarchs and lord of Salamanca, who gave it to the University. Its walls are also decorated by 17th century tapestries from Brussels and a portrait of Carlos IV which bears formal similarities to the work of Goya. The academic year is commenced in this classroom.

University Paraninfo

« ***As we were saying yesterday.*** »
Friar Luis de León.

Friar Luis de León Classroom

THE UNIVERSITY STAIRWAY

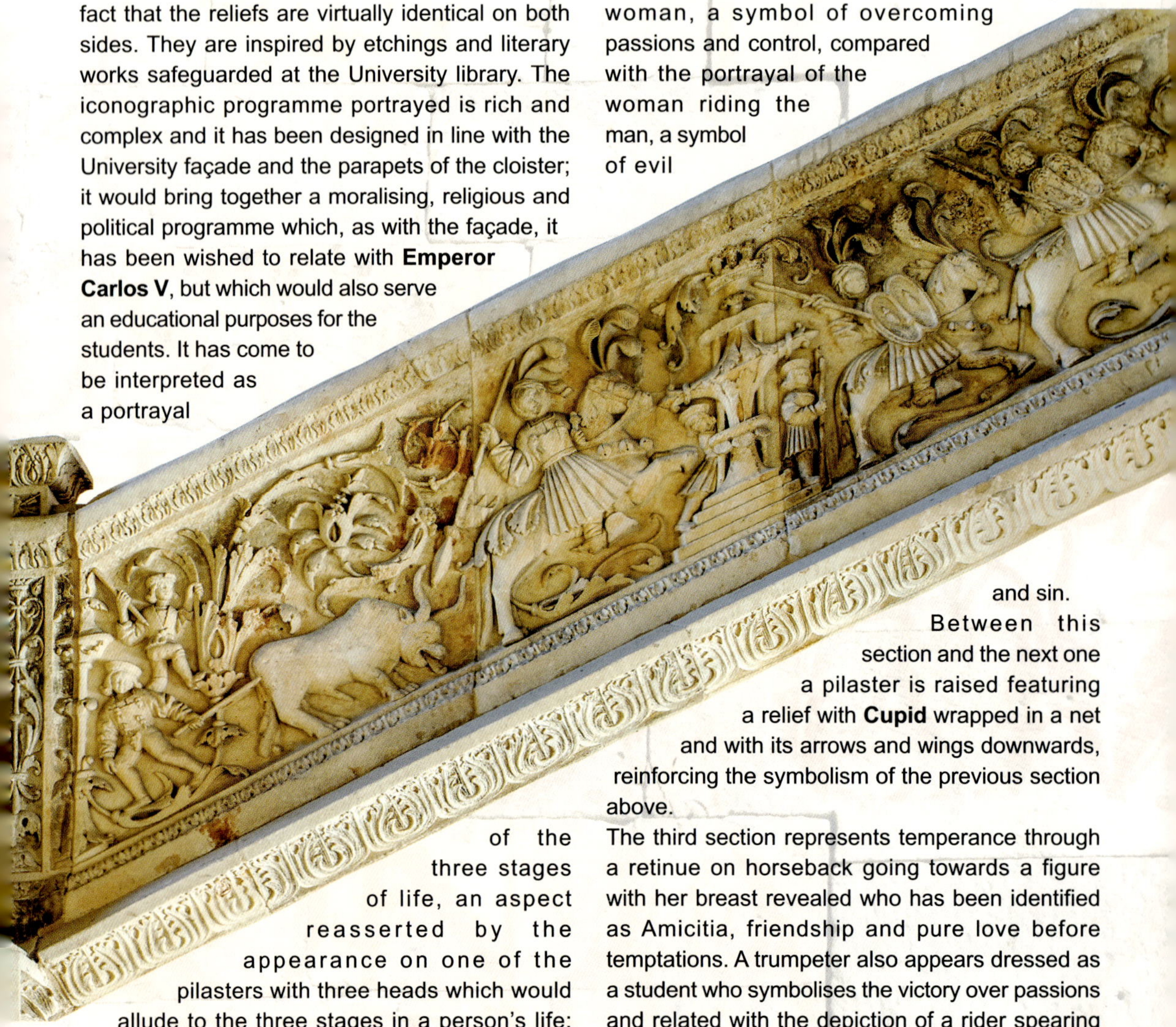

The **steps** were carried out during the first quarter of the 16th century by an unknown author or authors as some people say that each of its sides may have been carried out by a different artist who copied the previous one in view of the fact that the reliefs are virtually identical on both sides. They are inspired by etchings and literary works safeguarded at the University library. The iconographic programme portrayed is rich and complex and it has been designed in line with the University façade and the parapets of the cloister; it would bring together a moralising, religious and political programme which, as with the façade, it has been wished to relate with **Emperor Carlos V**, but which would also serve an educational purposes for the students. It has come to be interpreted as a portrayal of the three stages of life, an aspect reasserted by the appearance on one of the pilasters with three heads which would allude to the three stages in a person's life: youth, maturity and old age; furthermore, on the first pilaster a pilgrim is depicted, looking upwards, reinforcing this theory of life as a path with different stages in which various trials have to be overcome; this pilgrim also refers to the religious nature it has been wished to lend to the stairway.

The first section of the stairway would provide a representation of youth through a **Moorish-style dance**, a festive atmosphere with a jester, a courtesan and minstrels; it alludes to the hazards and temptations faced by the young student and to the unleashing of passions.

The second section provides a depiction of maturity through a juxtaposition of good and evil with the representation of a man riding a woman, a symbol of overcoming passions and control, compared with the portrayal of the woman riding the man, a symbol of evil and sin.

Between this section and the next one a pilaster is raised featuring a relief with **Cupid** wrapped in a net and with its arrows and wings downwards, reinforcing the symbolism of the previous section above.

The third section represents temperance through a retinue on horseback going towards a figure with her breast revealed who has been identified as Amicitia, friendship and pure love before temptations. A trumpeter also appears dressed as a student who symbolises the victory over passions and related with the depiction of a rider spearing a bull, the symbol of passions; in this regard it is worth highlighting that the students were celebrating the obtaining of their degree with a bullfight so it would also be referring to the end of student life.

Above: third section of the stairway, detail of a speared bull and a retinue on horseback

View of the stairwell

THE ENIGMAS

The library gallery was the first to be built on an upper floor and here we can find parapets from the **16th century** which, for a long time, proved indecipherable which is why they are known as the enigmas; they comprise 7 paintings, 5 of them taken from ***The Dream of Poliphilus***, a literary work from the late 15th century written by **Francisco Colonna**. Although they are endowed with a slightly complex iconographic programme, they do serve a moralizing and educational purpose for students. They occur from left to right:

The **first enigma** is made up of two reliefs alluding to moderation; in one two children appear holding a sphere with the slogan: Virtue is the happy medium. In the other, a woman sitting with a tortoise in one hand and in the other two wings alluding to the slogan *Festina lente*.

The **second enigma** also shows two scenes; the first features Cupid firing arrows into the sky, in the second it appears between two characters holding a placard which reads *Nemo* (no-one) in reply to

First enigma, Virtue is the happy medium

Second enigma: Cupid and Who can escape from love

the inscription of the previous scene: *Who can escape (from love)?*

The **third enigma** depicts two scenes separated by a lampstand; on one side, an anthropomorphous monster with three heads and the inscription *He will wound, destroy and laugh* in allusion to the passage of time; on the other side an eagle, the symbol of Jupiter, over the giants Typheus and Briareos and the slogan thou wilt neither say nor do anything against *Minerva* (the Goddess of Wisdom), so it may symbolise the University overcoming the giants of ignorance.

Numerous objects appear in the f**ourth enigma** and a very long inscription which refers to the fact that man must offer up the fruits of his labours to God, endowed with a clearly moralising purpose.

The **fifth enigma** contains two medallions; the first features the scales of justice accompanied on one side by a dog and on the other by a serpent with a slogan referring to the righteousness of Justice which won't be influenced either by friendliness (the dog) nor hatred (the serpent). On the other medallion, in the centre, there is a Cadeceus with serpents entwined and on the sides ants turning into elephants and elephants turning into ants under the slogan: *Peace and harmony make small things grow, disharmony makes great things decay.*

In the **sixth one** we can once again find a reference to *Festina lente* (hurry up slowly) with the dolphins, the symbol of speed and the anchors, the symbol of stillness.

The **seventh** and final **enigma** depicts eight serpents entwined in two groups, biting each other's tails in reference to the cyclical passage of time.

LOCATION OF THE ENIGMAS
WESTERN OR ENIGMAS WALLS

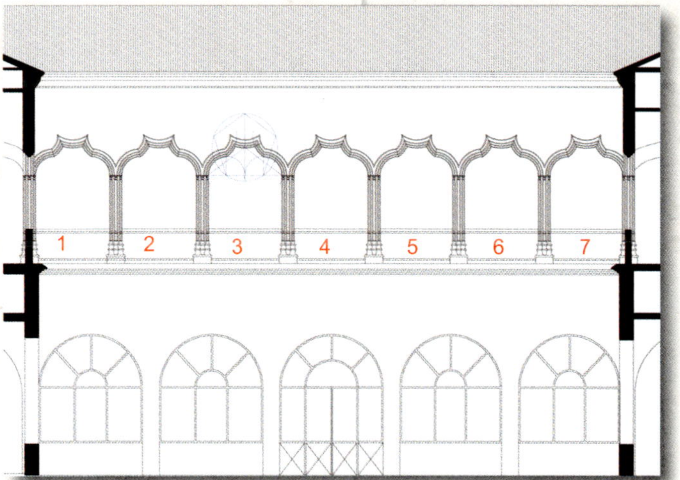

THE UNIVERSITY LIBRARY

We are now viewing the third **University Library**; there used to be an early library situated at the current university chapel which was demolished to lend greater height to the religious space; this early library contained the murals known as **The Sky of Salamanca**, now on display in a room at the University College courtyard. After the disappearance of this early library, a new location had to be sought for this element was so vital to university teaching; it was decided to situate it on the second floor above the western wing classrooms to which end the building structure had to be strengthened in this area in the early 16th century. Since the beginning this new library suffered structural problems which led to its ruin in 1664. The architects who worked on this second library include **Juan de Álava**, **Juan de Badajoz** or **Pedro de Larrea**. Of the former space the richly decorated access door and its grate were retained.

The University was to wait almost a century for its new library owing to a lack of economic resources; the author of this third and final library was **Andrés García de Quiñones**. The current library is rectangular in shape, covered by a barrel vault with lunettes and transverse arches which rest on corbels with angels. The library collections are organised on double-storey shelves whose design can be put down to **Manuel de Larra Churriguera**. The coats-of-arms over the doors belong to **Simón Gavilán Tomé**. Worthy of special mention, at the four corners, are the sandstone sculptures with depictions of Mercury, Chance, Abundance and Fortune.

Its important collections include works like ***El Libro del Buen Amor*** (The Book of Good Love) or the ***Liber Canticorum***, a Mozarab manuscript. It also has a fine collection of globes.

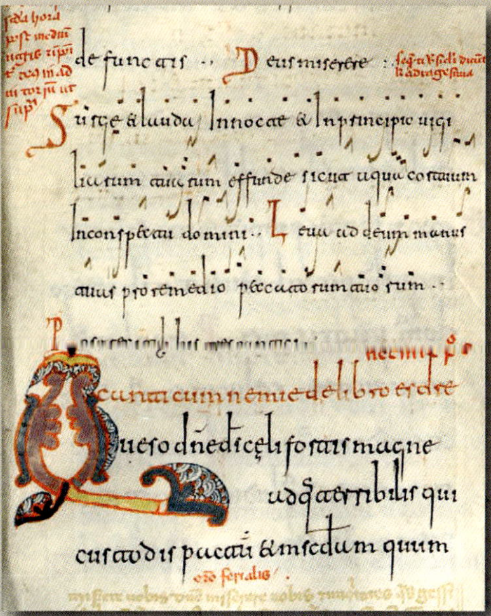

Musical notation in the Liber canticorum et horarum (Ms. 2668). Historical Library of the University of Salamanca

HOUSE-MUSEUM OF MIGUEL DE UNAMUNO$_{9}$

A former rectory house, this **mid-18th century** building currently accommodates the museum of the famed writer who lived here from 1900 until 1914, during which period he was the **University Rector**. Inside there are household items, furniture, manuscripts, photos and books of the author, seeking to recreate his way of life.

The building is located alongside the old University and it was erected between 1758 and 1761 to accommodate the cloisters' room; the work was commissioned to the architect **Andrés García de Quiñones** whose mark is clear in the window frames, similar to those to be found at the Town Hall on the Main Square[116], also his work. Worthy of special mention are the gateway where the specific characteristics of the Baroque are evident with winding lines and hefty volumes. The most appealing thing about the building is perhaps the university coat-of-arms situated at the corner of the property, seeking that theatricality so typical of the Baroque.

ST.ISIDORE AND ST.PELAYUS$_{14}$

According to tradition, the Romanesque temple situated here was built on the occasion of moving the relics of San Isidoro from Seville to León. Nothing remains of that original temple as it suffered a fire in 1889. It was the seat of the Plateros de San Eloy Brotherhood and accommodated the parish church of St.Pelayus when the latter was demolished in the 17th century on the occasion of the construction of the Colegio Real de la Compañía de Jesús[84]; furthermore, it was the place of burial of illustrious figures, some of which were moved to the Old Cathedral[18] when the church fell into disuse. After many ups and downs, it is now a University lecture room. It preserves the **16th century gateway** which was clearly influenced by Rodrigo Gil de Hontañón, but the most striking thing is the window of the former side chapel to the main chapel, a **18th century** work whose arrangement is reminiscent of the works of Joaquín de Churriguera.

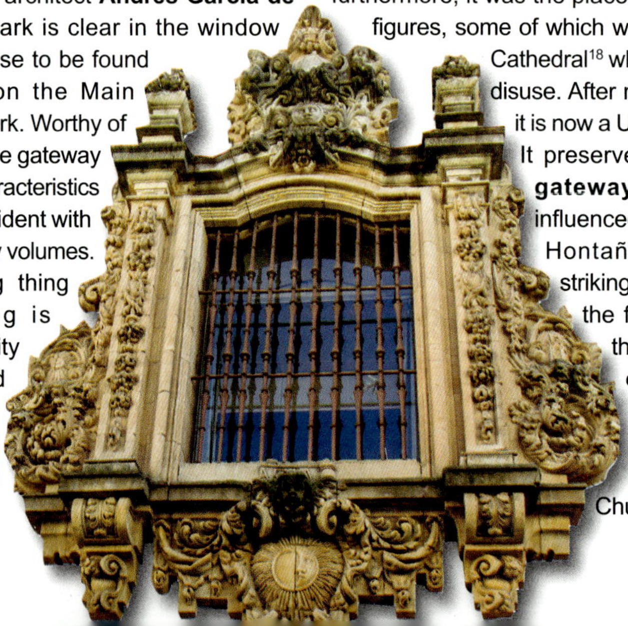

CALLE LIBREROS

CASA DE LAS CONCHAS 16

This is one of the most representative buildings in Salamanca whose construction was ordered during the reign of the Catholic Monarchs by Rodrigo Arias Maldonado, combining a series of approaches and resources which leave it half way between the Hispano-Flemish style and the Renaissance, characteristic of this age of transition from the Middle Ages to the Modern Age.

Although it is amazing to behold, in the past it must have been more so as it reached us with a fair remodelling work and added items; hence, only the main façade is original as the side is a 18th century reconstruction, at which time it underwent many repairs owing to its poor foundations which brought about the disappearance of the side towers and the façade paving. These latest remodelling works were undertaken by **Simón Gavilán Tomé** in the last third of the 18th century.

The most appealing thing about this building is undoubtedly its **more than 300 shells** adorning its façade and lending it its name; the presence of this decorative element

LA COMPAÑÍA

repeated to the point of society is easy to explain; firstly, the construction promoter was a Knight of the Order of St.James and, on the other hand, the house was erected on the grounds of the betrothal of a Maldonado with a lady from the Pimentel house which bears on its coat-of-arms, amongst other things, the shell which can be interpreted as a display of love. In any case, it was extremely frequent at the time to adorn the palace façades with motifs which would be repeated throughout the whole wall such as diamond tips. According to tradition, behind each of them there lay treasure.

Original iron doors

The gateway appears displaced from the central axis, not coinciding in the interior with access to the courtyard, the device employed to safeguard the intimacy of the house and its residents whose forefathers could be gleaned from the elbowed entrances of the Islamic world. Over the gateway a mixtilinear tympanum opens out bearing the coat-of-arms of the Maldonado flanked by two lions. The spans of the lower storey are closed by **iron**

doors whilst those of the main storey boast various designs, the common element being the coats-of-arms of both families: the Maldonado and the Pimentel.

In the interior there lies the double-height **courtyard**; the lower storey opens out into mixtilinear arches which served as the model for subsequent constructions such as the Schools' Courtyard[56] or the Salina Palace[126]. On the upper floor the columns stand out which support the arches made of Carrara marble; between them the parapets bear various plaiting and honeycomb designs.

It currently accommodates the outbuildings of the Public Library.

VIRTUAL TOUR GIGA

THE CLERGY ~ PONTIFICAL UNIVERSITY

The Royal College was founded in **1611** by **Queen Margaret of Austria**, the wife of Felipe III. The Jesuits already had a college in the city located alongside the Archbishop Fonseca School[104], but it was thought that they should have a larger one in the city centre, closer to the University, hence its current siting, though its construction took more than **150 years** as problems arose in the acquisition of plots as to undertake the project the St. Pelayus church and the St.Catherine hermitage were pulled down as were numerous houses combined with the lack of economic resources.

The works were started in **1617** and the features were provided by the royal architect **Juan Gómez de Mora**; essentially, the new building shares with the old university its sober, simple appearance though it is larger in size.

After the expulsion of the Jesuits from Spain in 1767, the church college began to be run by the **Clergy of St.Marks**, being known from that time on by the popular name of La Clerecía (The Clergy). Since 1940 the college premises have been occupied by the **Pontifical University**.

The **Royal College** is accessed via a decorated **gateway** carried out by **Jerónimo García de Quiñones** with modifications after the features afforded by his father Andrés. It was built after the expulsion of the Jesuits when this area already belonged to the Conciliar Seminary. Worthy of note there its framed decoration and the accumulation of architectonic elements.

The College as whole is endowed with U-shaped planimetry, with the long sides comprising the **pavilions** crowned by their characteristic galleries. This part of the building is austere in terms of its sober, majestic decoration; worked was carried out on the pavilions by Juan de Setién, Joaquín de Churriguera and Andrés García de Quiñones.

At the centre of the whole there lies the **Studies Cloister**, one of the best Spanish Baroque works. It comprises three upper floors; the first two are joined by giant columns and the last one corresponds to an attic which finishes off the lower columns with piers and pinnacles. The second floor accommodates the decoration in canted arches which boast balconies and oval oculis whose frames are bedecked in castles, lions, fruits and coats-of-arms. This work is in perfect harmony, combining verticality and dynamism with the interplays of volumes and the chiaroscuro without excessive decoration and endowed with a rhythmic sense of composition. The cloister interior

GROUND PLAN OF THE CONSTRUCTION STAGES
General features of the work of Juan Gómez de Mora

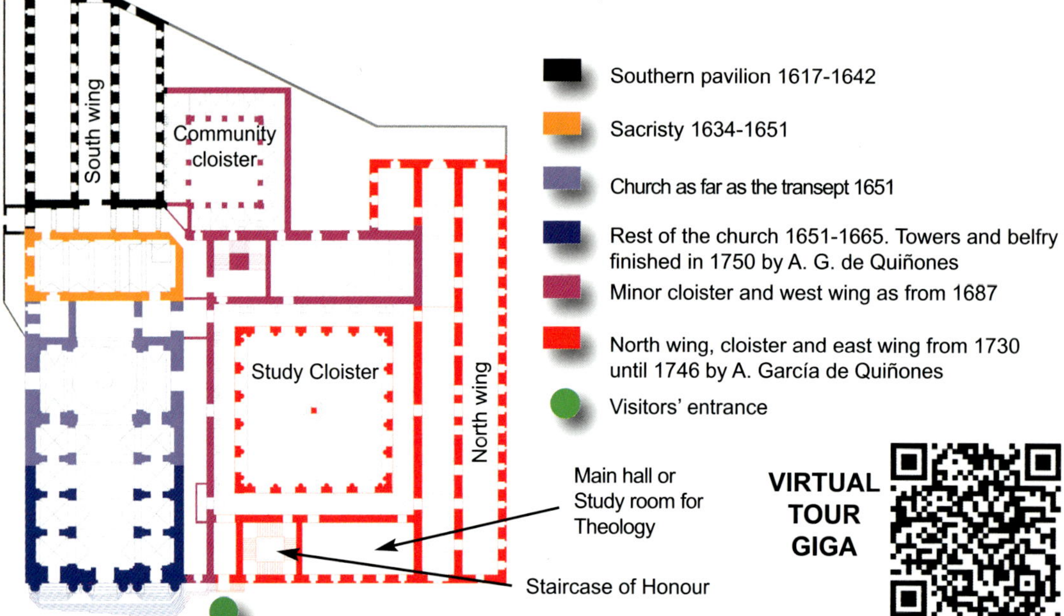

VIRTUAL TOUR GIGA

accommodates an exhibition of **canvases** recounting the life of St.Ignatius of Loyola carried out by the painter Sebastiano Conca.

On one of the wings of the cloister there lies the magnificent **stairway** undertaken by **Andrés García de Quiñones**, inspired by the **Escalera de Soto** at St.Stephen's Convent[146]; the same as the former, it is a cantilevered staircase whose ramps are adorned by panels housing bulging motifs. The vault is covered by plasterwork and stucco providing depictions of the coats-of-arms of the founding monarchs.

SCALA COELI

This is the exhibition organised around the visit to the two church towers and which seeks to make visitable rooms which were unknown until recently as well as showing magnificent views over the City and the Cathedrals of Salamanca.

In the stands scale models can be seen and documentation with the history behind the construction and its craftsmen.

Partial view of the route via the towers

Claustro de Estudios

On the north wing of the cloister is the **Paraninfo** or **Theology** study rooms, undertaken by **Andrés García de Quiñones** in the first half of the 18th century; worthy of special mention is the vault covered in plasterwork and stucco whose iconographic programme can be put down to **Father Lino Franco** and it is devoted to the exaltation of Theology. On it there is a depiction of the Doctors of the Latin Church and the four great doctors of the Company of Jesus: Francisco Suárez, Luis de Molina, Gabriel Vázquez and Gregorio de Valencia; there are also portrayals of St.Thomas Aquinas and St.Luis de Gonzaga as well as allegories of the virtues and inscriptions. On the large façade, there are ample canvases portraying themes of the Company of Jesus with the apparition of St.Ignatius of Loyola about the allegory of Wisdom, sending out its members to evangelise the world; on the other there is a depiction of the Council of Trent with Father Diego Laínez setting out his theses. Both canvases were painted by Juan Ruiz Soriano.

Paraninfo or Theology study rooms

Nec me capit angulus unus.

THE CHURCH

The **church** is laid up in Jesuit style with one nave and side chapels connected to each other. The features were provided by Juan Gómez de Mora, but from 1647 onwards Pedro Mato took up the Works, undertaking to carry out the second body of the façade, the vaults and the dome; the action taken by the latter architect resulted in a Baroquisation of the initial project, far more sober and closer to classicism. However, the culmination of the Baroque arrived with **Andrés García de Quiñones** who took charge of the Works in **1749** and added the two towers and the central belfry which are richly decorated. The **façade** is characterised by its grandiose nature, in which the giant columns stand out and the bulging coats-of-arms and reliefs. Over the entrance door, inside a niche, there lies the image of **St.Ignatius of Loyola** carried out by **Juan Rodríguez** who was also the craftsman behind the reliefs of the gateways of the New Cathedral[26]. However, it is undoubtedly the towers and the belfry that have lent personality to the façade and its characteristic profile which, cut against the skyline, competes with the Cathedrals; its breath-taking nature gains is further enhanced when observed from the ascent of Calle Palominos. Both the **towers** and the belfry appear intricately decorated; the former with sculptures of the Fathers of the Church, the Virtues and Jesuit saints. The belfry has a major relief with the theme of the Coming of the Holy Spirit and it crowned by a sculpture of the Virgin flanked by the monarchs Felipe III and his wife Margaret of Austria, the founders of the College.

Its **dome** is also impressive, greatly remodelled over time as well as its structural problems we need to add the damage suffered during the Lisbon earthquake of 1775, requiring parts of the windows of the drum to be shut off, which have been reopened, and to girth it. In the interior the decoration bedecks the penditives where the monarchs' coats-of-arms appear.

The **interior** of the church is sober and simple, cold in appearance. The upper part of the nave is surrounded by the stands; the decoration is limited to the large grooved pilasters and the cornice.

Neither does the **main retable** boast any decorative fanfare, standing out for its sobriety and simplicity despite the presence of the Salomonic columns which lend it a certain dynamism. The sculptures were carried out by Juan Rodríguez who had already worked on the façade; on the intercolumns there are sculptures of the Four Fathers of the Church and the central part depicts the Coming of the Holy Spirit or Pentecost, over which there is a large placard with the anagram of the Company of Jesus. On the latter body the scene of St.Ignatius of Loyola is played out, writing the exercises of the Order inspired by the Virgin and the Trinity; the four evangelists sit on the cornice.

The transept retables are dedicated to St.Ignatius of Loyola and St. Francis Javier; the sculptures and reliefs were carried out by the same master of the main retable.

This church is the seat of various Easter Week brotherhoods, highlighting the **University Brotherhood** which takes out the Holy Christ the Light and Our Lady of Wisdom out on procession, a 17th century work by **Esteban de Rueda** and the Brotherhood of Our Father Jesus Whipped which bears the image of a high-quality Ecce Homo carried out by **Luis Salvador Carmona** in the 18th century; both images are located in the altars of the side chapels.

THE ST.BENEDICT'S CORNER

ST.BENEDICT'S CHURCH[17]

During the Late Middle Ages this church was one of the most important parish churches as it united around it one of the noble groups, the benitinos, confronted as they held sway over the city. The building we can see today is not the Romanesque one, that was replaced in 1506 by order of **archbishop Alonso de Fonseca**, the patriarch of Alejandría; it is a late Gothic building with just one nave and a polygonal header.

On the exterior it has sturdy buttresses and an interesting Hispano-Flemish gateway on the theme of the Annunciation presided over by God the Father; at the buttresses of the building, as well as at the gateway, you will see the noble coats-of-arms of the Archbishop.

The interior serves a funeral purpose as there are many niches opened to accommodate tombs; the majority belong to the Maldonado family which used is a family pantheon, with the most interesting ones being the two tombs of the main chapel with the recumbent figures carried out in alabaster.

MADRE DE DIOS CONVENT[20]

This convent was founded in the mid-**16th century** by Francisco de Loarte and his wife María de Castro; it occupies the former house of Pedro Maldonado, the head of the Revolt of the Comuneros who as executed in Simancas after the defeat at Villalar. As occurred to other families, Carlos V ordered their goods to be confiscated, to chop up the noble coat-of-arms and seal up their doors as a punishment for the uprising. In its interior you can still see the old layout of the house in the early 16th century. During the transformation works and the adaptation to the new use, the architect **Juan de Nates Naveda** was involved. Worthy of special mention are the church Gateway flanked by pilasters with Ionian chapters; around the cornice we can see a Gothic-style frieze, a remnant of the old house.

MALDONADO HOUSE[19]

This house belonged to the servant to Archbishop Fonseca, Diego Maldonado Rivas; it was crafted by **Juan de Álava** who at that time was working in Salamanca where he also carried out the Archbishop Fonseca College[103], St.Stephen's Convent[138] and the nearby house of the Deaths[112] with which it shares stylistic devices.

The **façade** is now fully **Renaissance** and the Gothic conception has been abandoned; it has a lintelled gateway and over it a window opens out framed by two pilasters decorated by friezes. Over the window there lies the coats-of-arms of the Maldonado and on the sides, the coats-of-arms of the Medrano and the Morille, families with which they were related. Crowning the whole is the coat-of-arms of Archbishop Fonseca; all supported by putty in various positions.

SOLÍS HOUSE [18]

Located at St. Benedict's square, alongside the previous one, the house has certain similarities to the latter in terms of the conception of its façade. Dated at around **1540**, it opens out with a lintelled door framed by two columns in the form of an alfiz which support an entablature decorated by friezes; underneath it, the coats-of-arms with the depictions of various noble families from Salamanca with which the Solís were related. On the entablature there is an adorned window on which the families' coat-of-arms lies. Finishing off the façade, a balustrade of whitewashed circles which would probably have been a gallery.

Maldonado house in the background, on the right the Solís House

PALACIO DE MONTERREY 22

This striking building was constructed by the 3rd Count of Monterrey, Alonso de Acevedo y Zúñiga who in **1539** commissioned **Pedro de Ibarra**, inter alia, to follow the features provided by **Rodrigo Gil de Hontañón** and **Friar Martín de Santiago**. The project was never fully completed as it was going to consist of four wings around a courtyard with four towers at each angle, though there are sources who suggest the construction of a further four towers in the centre of each wing as there are today; this aspect remains unclear but it would seem that what we can see today corresponds to an initial stage and a subsequent extension towards the site of the Franciscans. In any case, this Plateresque building has served as the model for subsequent palaces, including, in the 19th century in those styles called revival and eclecticism stylistic devices began to be copied again; to see its influences, look no further than in the city itself where we can find the Solís Palace[159], rebuilt in 1934 by Telefónica following the parameters of the Monterrey palace.

Restricting ourselves to the building itself, it should be pointed out that its extensive decoration was changed in the 19th century when it was restored and the ends and fronts of the windows on the main floor were removed owing to their poor condition due to erosion which are preserved on one side of the tower. The most characteristic thing about this building are its towers and the **high gallery**; its **finishes** are richly decorated with tracery where figures and candle holders follow each other, the decoration of coats-of-arms of the house lineages, cupids and medallions occupy other spaces, lending the building fine decoration near the filigree; the decorative programme is so perfect that even the chimneys have been decorated, apparel losing is functionality to become a decorative, decorated element.

The building currently belongs to the heritage of the Duchy of Alba House.

The original location of the convent was situated on the banks of the river, but the already well-known St. Polycarpus flood of 1626 obliged the monks to abandon the place and find another residence; it was then that **Manuel de Zúñiga y Fonseca**, the 7th count of Monterrey and the Viceroy of Naples, became interested in the matter and paid for a new convent and church alongside his palace which would later serve as a family pantheon.

The convent was designed by **Juan Gómez de Mora** and it is very sober, including its façade, carried out by **Joaquín de Churriguera**. However, the features of the new church were commissioned to the Italian architect Bartolomeo Pichiatti and revised by **Cósimo Fanzago** who took over the work as well as the design of the altars and pulpit.

The church type used is purely **Italian**, both in terms of the ground plan and the elevation. It has a Latin cross with just one nave with two side chapels and a portico on each side of the main gateway. As regards its **façade**, this is of the Vignola type with two bodies crowned by a triangular front and joined by two buttresses; the gateway is very original as it is flanked by pilasters whose shafts are made from marble blocks sculpted into a diamond tip.

On the exterior its **dome** also stands out, not in the slightest characteristic of Salamanca as it is a Madrid model; the first dome designed underwent construction problems and it caved in which is why a new lighter one was designed, following the advice of **Friar Lorenzo de San Nicolás**, an important architect and author of the Court. The new dome is chamfered and it has been made of wood and lined in slate on the exterior and with plaster on the interior which is why it barely weighs anything at all. This type of dome is very common in the Baroque buildings of Madrid.

The interior of the church is very austere and sober, though it does hide important artistic works by the great masters of the time; the most prominent of all is the canvas presiding over the main retable, a work by **José de Ribera**. This painting depicts the theme of the **Assumption of the Virgin**; it is large in scale and in it the artist has abandoned the chiaroscuro which characterised it to experiment with light. The centre of the composition is the Virgin who appears surrounded by angels weighing their symbols, above are the depictions of God the

The Immaculate Conception by José de Ribera

Father and the Holy Spirit accompanied by a choir of angels. Golden tones and bright colours prevail. With this painting, Ribera created a new model of Mary representation which was later followed by other Baroque painters, highlighting Murillo, very prolific and famous for portrayals of this type.

At the church we find further works by Ribera on the retables of the transept, but smaller in scale. However, our attention is drawn by the **tombs of the donors** with the praying statues, situated in the niches of the main chapel. Also worthy of special mention are the **pulpit** by Fanzago, richly decorated by coloured marbles and supported on a black marble eagle with its wings spread out; underneath there is a large coat-of-arms of the patron carried out in the same materials.

CHAPEL OF THE THIRD ORDER OF ST.FRANCIS[23]

This chapel, also called the Capuchinos chapel, is located on the plot which occupied one of the side apses of the **St.Francis and Royal Convent** founded in the 13th century; as regards said convent endowed with large dimensions which counted on the patronage of Prince Fadrique, the son of Fernando III as well as the predilection of the noble families of the city to be buried there, nothing remains except part of its head. Of the three polygonal apses, two remained standing, visible from calle Ancha and they serve as a good example of 14th century **pure Gothic**. The rest of the convent has gradually disappeared with the French occupation and the disentailments, with its cloister remaining in the former Holy Trinity hospital[156].

The **chapel** was built in the mid-**18th century**, following the features of **Andrés García de**

Quiñones and **Simón Gavilán Tomé**. The building consists of one sole nave with niche chapels on the side ones. Architecturally speaking, the interior is very sober with a design similar to Neoclassicism, but the intervention of Simón Gavilán Tomé in the construction of the main chapel and the retables, lent it a very heavy Baroque appearance. These are very interesting as, in addition to their typical Baroque decoration based on garden design, they stand out for the fact they were made of sandstone from Villamayor. The centre of the main retable is occupied by a Crucified Christ that at first sight recalls the work of the Sevillian Martínez Montañés, but its author was Bernardo Pérez de Robles who became familiar with the work of the former during his stay in Peru.

On the exterior the **gateway** reveals a more serene Baroque style, except in the dynamic moulding of the door; in the centre of the upper part, in a shell-shaped niche there lies the image of St. Louis King flanked by two figures on volutes, the work of Simón Gavilán Tomé and Gregorio Carnicero.

LOCATION OF THE CHAPEL AND THE ORIGINAL APSES

Remains of the original apses.

GIGA

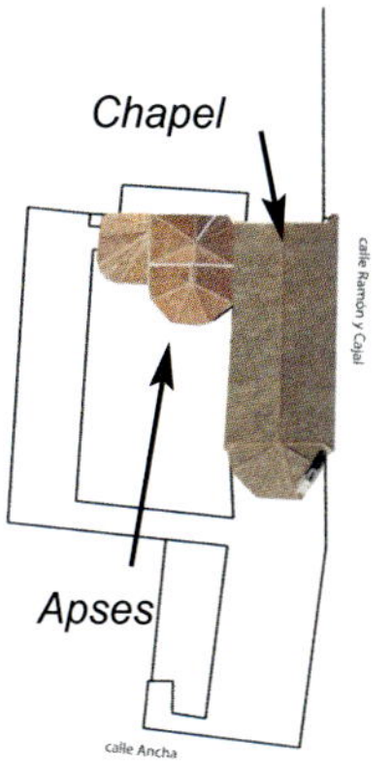

ARCHBISHOP FONSECA UNIVERSITY COLLEGE[24]

This is one of the four Universities which existed in Salamanca, of which only this one and that the Anaya College[38] remain as the Cuenca and of Oviedo College have disappeared. Their name derives from their founder and promoter **Archbishop Alonso de Fonseca**, although it is also known as the Santiago el Cebedeo College, to whom it is dedicated, or the Irish college as they occupied it almost continuously from 1827 to 1936.

As we have said, this foundation can be put down to Alonso de Fonseca, Archbishop of Santiago and Toledo, born in Salamanca and the son of the Patriarch of Alexandria. He took higher courses at Salamanca University and carried out great patronage work in the three cities he was linked to, keeping in contact with the most important architects and artists of the time such as **Diego de Siloé**, **Juan de Álava** or **Alonso de Covarrubias**. Of all his patronages, this was his favourite one as he was buried there and he bequeathed all of his goods and his library to it. The creation of the centre arose to provide accommodation to poor students and clerics, but over time it lost its original purpose and became an elitist centre.

The building started to be constructed in the first third of the **16th century**, with **Juan de Álava** serving as the commissioned architect. The original project was in quadrangular shape, being changed when the arms of the transept of the chapel and the presbytery were extended. Subsequently, the works were taken over by **Diego de Siloé** and **Fernán Pérez de Oliva** who lent the features to the gateway and the courtyard.

During the successive extensions and remodelling work that the building underwent, important architects from all ages were involved such as **Rodrigo Gil de Hontañón** carrying out the extension of the chapel and the construction of the annexed inn in the mid-16th century; the latter was replaced with a new one in the 17th century, commenced by **Juan de Setién Güemes**, and where the work of **Joaquín y Alberto Churriguera** and **Andrés García de Quiñones** is documented.

CHRONOLOGY OF THE BUILDING AND AUTHORS

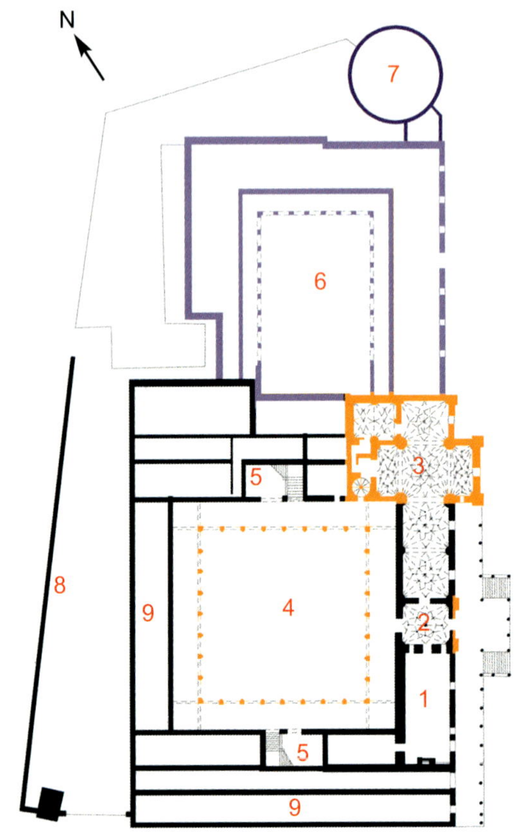

Juan de Álava, first third of the 16th century

Diego de Siloé and Fernán Pérez de Oliva, first third of the 16th century

Juan de Setién Güemes and Joaquín and Alberto de Churriguera, 17th century

Former anatomic theatre, Lesmes Gavilán1802 and modern extensions in the 21st century

1 Former main study room
2 Hallway
3 Chapel
4 Cloister
5 Stairways
6 Inn
7 Former anatomic theatre
8 Rests of the city wall
9 Cubicles or cells for accommodation

In the 18th century **Juan de Sagarvinaga** was working on the building. In the early 19th century a building was constructed with a central ground plan which served as an anatomic theatre of the Military Hospital set up during the Napoleonic occupation and which later led to the Faculty of Medicine being set up in the inn.

The most important elements of the building are, in addition to the gateway, the hallway, the chapel, the courtyard and the two stairways whose combining element was the coats-of-arms of the promoter.

The **gateway** was designed by **Diego de Siloé** and **Fernán Pérez de Oliva** which endowed it with totally Renaissance iconography, though still with some Plateresque reminiscences such as the friezes in the intercolumns, the heraldic medallions, cornucopias or the niches squinch which feature depictions of **St. Ildefonso** and **St. Augustine**, although they have also been interpreted as Archbishop Fonseca himself and his father. The upper part features a medallion with the image of **St. James the Moor-slayer** flanked by two winged angels with a human trunk and a vegetal lower part; completing the iconography, over the door there are three scallops with the symbols of the pilgrim. The apostle relief is very dynamic and denotes artistic expertise. During the course of the façade we can find windows which repeat the same decorative solutions as the gateway.

In the interior of the building we are received by the **hallway** which leads on to the chapel, the courtyard and other outbuildings. It is covered by a starred vault whose ribs extend along the walls as far as the ground, a characteristic specific to **Juan de Álava**. In the key room the **Fonseca coat-of-arms** appears again, also present in the four walls of the room, supported by winged putties.

On the right the gateway of the richly decorated **chapel** appears, recalling the façade of the House of the Deaths[112], also undertaken by **Juan de Álava**. The friezes take up almost the whole space, decorating the architectonic elements. In the spandrels of the arch, two medallions with busts representing Adam and Eve. Crowning the whole, there is another Fonseca coat-of-arms held up by angels.

Hallway, vault and gateway of the chapel

The chapel originally had a rectangular shape, but when its initial purpose changed and it became a **funeral chapel** to house the remains of the Archbishop, the header was modified, totally altering the appearance of the original work carried out by Juan de Álava. In 1540, **Rodrigo Gil de Hontañón** undertook the works to extend the chapel, with the latter consisting of the implementation of a new header and transept which exceeded the cubic volumes of the exterior of the building. Worthy of special mention in this area are the intricate vaults which cover the new spaces.

The **retable** was commissioned to **Alonso de Berruguete**, a Castilian artist educated in Italy. In it he highlights, over all the other paintings and sculptures, the image of **Piety**, endowed with serene dramatism and inspired by the same image of Michelangelo in the Vatican.

The **courtyard** was designed by **Diego de Siloé** with the collaboration of Fernán Pérez de Oliva; it consists of two storeys whereof the lower one has amore classical appearance characteristic of the Renaissance, though the upper one still recalls the Plateresque models of Tuscan arches and baluster columns. On both storeys the spandrels of the arches are adorned in **medallions** on which there are effigies of busts of unknown characters which prevent the correct interpretation of its iconographic programme.

VERA CRUZ CHAPEL 25

The building we see today is the chapel of the former Vera Cruz Hospital which disappeared in the late 16th century; its existence has been documented since the mid- 13th century when the Congregation of the Penitence of Jesús was formed, responsible for looking after the patients at the Hospital.
In the mid-**16th century** remodelling work was carried out on the chapel and the gateway dates back to this time, carried out by **Martín Navarro**, a disciple of Rodrigo Gil de Hontañón; it is a simple gateway in which the niche stands out with its image of the **Immaculate Conception** carried out by **Sebastián Dávila**.
In around **1713** new works were carried out which affected the internal appearance of the chapel and which were also felt in the exterior. The person responsible for this remodelling was **Joaquín de Churriguera** who lent the building a purely Baroque appearance; the vaults and walls were covered in mouldings and the **retable** was also carried out, designed by the same artist and adorned by the classical repertoire of this artist: Salomonic columns, stipes and curtains, rounding off the decoration with angels. In the main there is an image of the **Immaculate Conception** carried out by **Gregorio Fernández** in the first quarter of the 17th century, whilst in the attic there is a Gloria of tremendous character with the elements of the Passion carried by angels. A chapel opens out alongside where you will find the image of **Our Lady of Sorrows**, carried out in the first half of the 18th century by **Felipe del Corral** and inspired by the Virgin of Anguishes by Juan de Juni; it is characterized by its contained dramatism, emphasised by the twisting of the body and the folds in the garment.
This Vera Cruz chapel undoubtedly takes the limelight during **Easter Week** as it is the point of departure for many processions in which some of the best sculptural groups from Salamanca can be seen, carried out by artists of the level of **Alejandro Carnicero**.

It was founded in the second half of the **15th century** by Mrs.Sancha Maldonado, the aunt of **Archbishop Fonseca**, who played an active part in the reconstruction works on the building at the end of the same century to convert it into his funeral chapel. His **tomb** was carried out in marble by **Diego de Siloé** and it is situated at the centre of the presbytery; the purpose of this space as a family pantheon is evident, representing the different lineages to which the promoter belonged on the coats-of-arms around the walls of the apse. The exterior conception of the latter indicates that **Juan de Álava** was working on it, an architect in the service of the Fonseca on other works.

At the base of the temple there lies the overlaid double choir stall; the choir loft is closed by a Gothic grate made in a similar way to those of the Conchas House[80].

HOUSE OF THE DEATHS[29]

We are now at the house with the most beautiful façade in Salamanca; it was made by **Juan de Álava** or Ibarra who built it as a family house in the first third of the **16th century**, at which time he was working both on the Cathedral[18] as well as on St.Stephen's Convent[138] and other works ordered by his protector Archbishop Fonseca, for whom he also worked on Santiago de Compostela.

Taking a look at his peculiar name, there are several legends surrounding it, some relating to the terrible deaths of its residents owing to romances, jealousy and revenge; it is thus said that before the current building existed, the house of the Manzano family was in its plot who had deposited in the wine cellar the corpses of their two children decapitated by María la Brava. Another legend recounts the murder of several men at the hands of a spiteful knight which not only ended in the death of his wife's lovers, but also her and his own deaths too. Basically, it is worth thinking that the name derives from the former name of the street where it is situated, now called Bordadores and formerly called Deaths or skulls street which, like corbels, support the pilasters of the windows on the upper storey. Special mention should undoubtedly be given to the façade, a summary of the artistic personality of

Juan de Álava; here we can find the usual repertoire of this architect: medallions with busts, a gateway endowed with a lintelled span finished off by a large lunette, canted, the frieze with fantasy figures, the heads of cherubins with wings running around the cornice; all clearly influenced by the **Italian Renaissance**.

On the frieze over the door there lies the coat-of-arms of the Ibarra family, held up by two Cupids on dolphins; a similar scene, but without the animals, awaits us on the balcony where there is also a medallion with the effigy of Alonso de Fonseca, the Archbishop of Santiago and the main protector of Juan de Álava. The whole scenario is framed by two pilasters decorated with friezes on which an arch rests, decorated by cherubins. Flanking it, two medallions with characters dressed roman style which have been identified, according to the various sources, as Caesar, Brutus, Augustus or Antonio. The characters on the other medallions to be found on the façade are unknown and they wear the fashion of the time. The façade of this house bears a series of enigmas and devices which bring it close to the **University**[52] on which it is believed the architect may also have worked.

POLICIA LOCAL
POLICIA LOCAL

ST. MARK[34]

To the north of the original medieval city, alongside the city wall, there lies this odd circular church about which there has been much speculation and it has even been said that it was erected as a defensive turret of the city wall.

This type of **centralised design** was very common in the Romanesque of the late 12th century and early 13th century as we can see in the not too distant church of Vera Cruz de Segovia or the Navarre ones of Eunate and Torres del Río, all connected with Military Orders.

GROUND PLAN OF ST. MARK'S

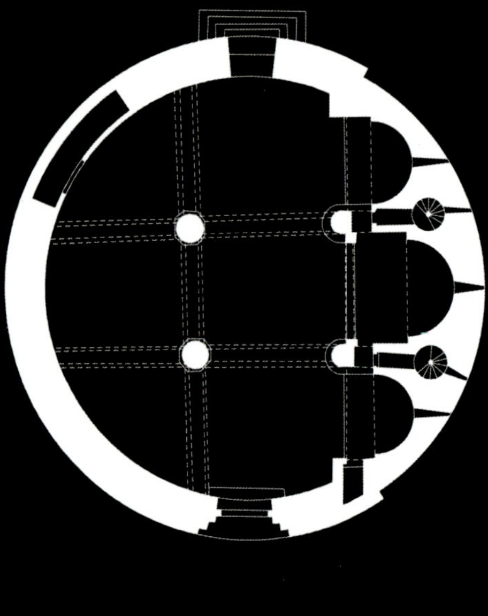

Inside of St. Mark's.

Although the exterior has this characteristic ground plan, upon entering inside we are surprised to find a **basilica temple** endowed with three naves and their corresponding semi-circular apses, perfectly adapted to the circular shape.

This building was a royal foundation, subsequently belonging, owing to a donation by Alfonso IX, to the clerics of Salamanca, thereafter receiving the name of the Royal Clergy of St. Mark; the clerks were here until 1767 when, after the expulsion of the Jesuits from the country, King Carlos III donated the College of the Society of Jesus to them and thereafter known by the popular name of The Clergy[84].

In 1967 restoration and consolidation works were undertaken which got rid of the Baroque additions, except for the belfry carried out in all likelihood by **Simón Gavilán Tomé**. Furthermore, there was the discovery of the altar tables and Gothic **murals** from the 14th century concealed behind the retables and which include scenes such as the Coronation of the Virgin, the Annunciation and St. Cristóbal accompanied by a further three saints. These paintings are stylistically related with those to be found at the church of San Sebastián de Toro, deriving from the St. Clare's convent and St. Martin's chapel at the Old Cathedral[18].

Painting of the Coronation of the Virgin

LA PLAZA MAYOR

VIRTUAL TOUR
GIGA

PANORAMIC VIEW IN FOLDER

Royal Pavilion, plaque with dedication

Salamanca's Main Square is one of the most majestic in Spain, though urbanistically speaking it has brought nothing new to the kind of squares being built at the time, what is for sure is that, aesthetically speaking, it attained levels that make it worthy of the title of the most beautiful of all for its harmony and balance.

For its construction the architect **Alberto de Churriguera** was called upon who, at that time, was working at the New Cathedral[26] of the city; this was his major work, though he never managed to complete it owing to various problems outside his control. The initiative derived from the chief magistrate Caballero y Llanes who was seeking not only to adorn, but also to offer a broader space for the market and the fairs as well as a place for festivities and the nerve centre of city activity. With this in mind, the former market place was used, back-to-back with the plot of the adjoining St. Martin's church[122], achieving a royal license on January 12th 1729 and starting the construction of

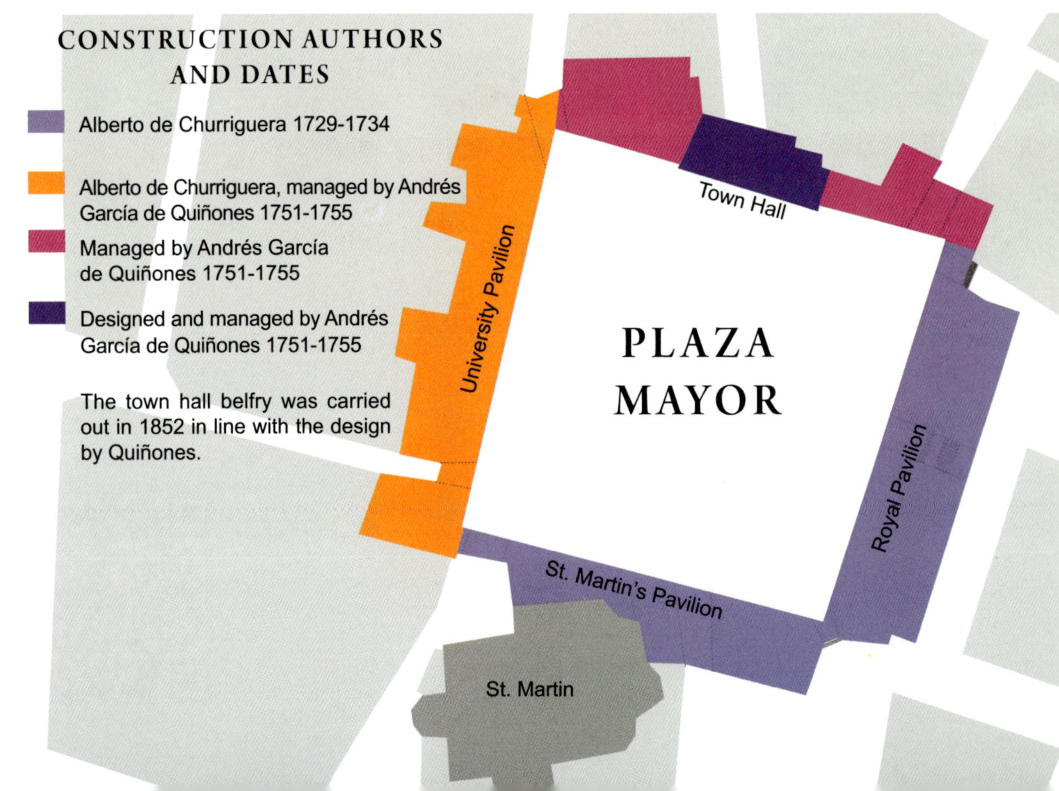

FARMACIA
Stradivarius

Count Fernán González

Rodrigo Díaz de Vivar, el Cid

Isabel and Fernando, the Catholic Monarchs

the new square in May of the same year, alongside the Royal Pavilion, from south to north.

Upon completing the works on the southern and eastern wings, problems arose with the private owners of the land who must have occupied the other two sides which is why the work came to a standstill for 15 years. After this stoppage, the works were resumed under orders of the Galician architect **Andrés García de Quiñones** who evened out the ground plan of the previous project as much as he could as this site is not totally even owing to the unevenness of the ground and the presence of St. Martin's church. The square was completed in 1755. Owing to the private nature of this last part of the Project, on the balconies of the first floor you will find the coats-of-arms of the Cathedral's Chapterhouse, the Royal Clergy of St.Mark, the St.Bartholomew School, the Caballeros Veinticuatro (Twenty Four knights), the Encomienda of St.John or the University.

Decoratively speaking, in addition to the aforementioned coats-of-arms, it was sought to lend some stylistic unity to the whole square which is why the differences are barely noticeable. Decorative medallions adorn the spandrels of all the arches with the busts of various characters, highlighting the monarchs of Spain in the Royal Pavilion from Alfonso IX to Carlos III; the heroes and conquerors are also represented on the St. Martin wing. The medallions of these two sides were carried out by **Alejandro Carnicero**, who was inspired, in the case of the monarchs, by some etchings. The characters stand out for the emphasis and arrogance typical of the Baroque. The iconographic programme remained unfinished and at present work is continuing on the medallions which remained empty with figures from culture and history related with the city.

The Town Hall, carried out by Andrés García de Quiñones himself and promoted by the Municipality, is grandiose in terms of its proportions and details; the planned towers were not built. It shows details of rock gardens and the application of stipes on the balconies of the upper body; there are also some floral touches which recall some retables planned by the same artist.

Antonio de Nebrija

Christopher Columbus

Alberto de Churriguera

Carlos I ó V

Felipe II

Carlos III

St. Fernando III, in the Royal Pavilion

Hernán Cortés

Francisco Pizarro

Diego Ponce de León

Southern Gateway

ST. MARTIN OF TOURS[47]

This **12th century** Romanesque temple was the largest in this style after the Old Cathedral[18] with which it competed in terms of importance, not only because of its size but also a bit was located in the busiest part of the city alongside the Market square.

Although, over time, it has undergone numerous remodelling Works, with elements being added and constructions backed onto it, it still retains part of its Romanesque past; hence, it consists of a basilica ground plan with three naves finished off by three semi-circular apses with no transept. It shows notable influences from the **Old Cathedral**[18] but also from other first line buildings within the Romanesque such as the **Collegiate Church of Toro** and **Zamora Cathedral**.

As we have indicated, its appearance has gradually varied over the centuries, hence in the **18th century** its vaults had to be redone by dint of the collapse of the original ones, with **Jerónimo García de Quiñones** taking charge of the project. The 16th century saw the construction of an atrium in Gil de Hontañón style and in the interior the choir stall was raised with a Plateresque stairway reminiscent of the University[42] one; in the 17th century it backed onto the base of the Carmen chapel, shutting off the Romanesque western gateway.

Of the three Romanesque gateways it had, only the northern one has been retained and its appearance betrays a direct link with the **Bishop's Gate of the Zamora Cathedral** in the lobulated arches of one of its archivolts. Above it appears the depiction of St. Martin on horseback, taking off his cloak to shelter a poor man; this scene can be seen again on the southern gateway.

CONSTRUCTION STAGES AT ST. MARTIN OF TOURS

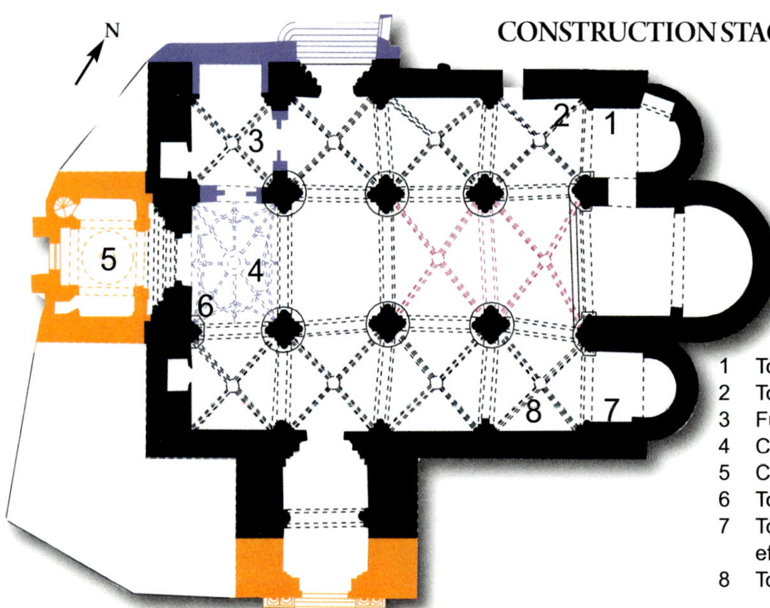

- Romanesque 12th to 13th centuries
- Gothic 15th century
- Late Gothic 16th century
- 18th century and 19th century, remaking of the vaults after their collapse.

1 Tomb of Diego Santisteban
2 Tomb of Roberto Santisteban
3 Funeral chapel of the Count of Grajal
4 Choir stall and Retrochoir
5 Carmen Chapel
6 Tomb of Antonio de Paz et al.
7 Tomb of Pedro Bernardo del Carpio and effigy of Andrés Santisteban
8 Tomb of the school graduate Luis Yánez

Northern Gateway

In the interior of the temple the condition can be observed in which the building ended up after the destruction of part of the original vaults; the pillars supporting them subsided as the lateral naves were unable to withstand the thrust of the latter and they had to be partially redone.

The most interesting aspects about the interior are the capitals of the naves in which the work of two different workshops can be distinguished between. Also present in the interior we can find a series of **15th century tombs** belonging to various characters, highlighting those of the Santiesteban family, which include that of Roberto Santiesteban with the posture of the sculpture of the deceased denoting similarities with the famous tomb of the Young Nobleman from Sigüenza.

In the main chapel there was, sometime previously,

Relief of St. Martin taking off his cloak to shelter a poor man

Former Romanesque gateway now inside the Carmen chapel

Detail of the archivolt, from left to right: Monster; February: pruning of the vine; March: people warming up.

a magnificent retable designed by Juan Gómez de Mora in the first quarter of the 17th century and which included sculptures and reliefs by Esteban de Rueda and Antonio de Paz; unfortunately, it disappeared in a fire in 1854.

Undoubtedly, sculpturally speaking, the western gateway is the most interesting one as on it, besides other depictions, we can find scenes which, though displaced, refer to a **farming calendar** which could be related with others nearby such as that of Saint Claudius of Olivares in Zamora or Saint Isidore of León. Between the scenes, we can make out, worthy of special mention is the one with a man pruning vines and others getting warm around the light of a fire.

PALACIO DE LA SALINA 64

This unique building began to be built in the first half of the **16th century**, following the design by **Rodrigo Gil de Hontañón**; its promoters were Rodrigo de Mejía and Mayor de Fonseca, though tradition attributes it to **archbishop Alonso de Fonseca** who was said to have ordered its construction to shelter his lover, rejected by nobles of Salamanca. The fact of the matter is that the palace postdates the death of the archbishop and he could thus not have been its promoter, but rather his niece Mrs.Mayor who bequeathed it to her second son Juan Alonso de Fonseca who had the concession of the salt pond of the city which is why it was called "salina" (the saltworks).

As far as the building is concerned we can find differences from others in the city itself; its façade opened by an extensive **loggia** is more characteristic of a public building than a residence, though there were precedents in the Italian palaces of the 16th century; we can also find similarities with the hallway of the Escuelas Menores[56] (schools) which may also be by the same author. Putting aside these reflections, the façade of this building is odd not only because of the high logia and its arches of different sizes, but rather also because it is decentred from the two upper floors, leading us to think about a different author or implementation date. Whatever the case may be, it is a beautiful façade whose upper gallery recalls the **Monterrey palace**[96] and whose medallions lead us to the cloister of the Archbishop Fonseca college[92]. These **medallions** bear depictions of female and male characters and those identified are the effigy of Cleopatra with the asp and that showing Mark Anthony.

Inside we can find an irregular-shaped **courtyard** whose oldest part is the front with a gallery of **mixtilinear arches** which recall the courtyard of the Escuelas Menores[56] (schools) and that of the Conchas House[80]; however, the most attractive aspect is undoubtedly the

corbel of the upper gallery where bare figure can be observed endowed with pronounced anatomy and gestuality, being supported on each other like **atlantes** which look like the carvings of the Las Dueñas[150] cloister; as tradition would have it, these figures represent the nobles of Salamanca who did not provide any aid to archbishop Fonseca who thus put them to ridicule in this way.
The building is currently occupied by the County Council.

Medallion of Cleopatra

TORRE DEL AIRE [50]

This is a **15th century** palace owned by the Castilla family, lords of Fermoselle; its name derives from its high tower which recalls the Italian Gothic palaces for its slenderness. Its features have changed greatly, retaining some mullioned windows framed by an alfiz with tracery decoration in its interior.

CALLE COMPAÑÍA

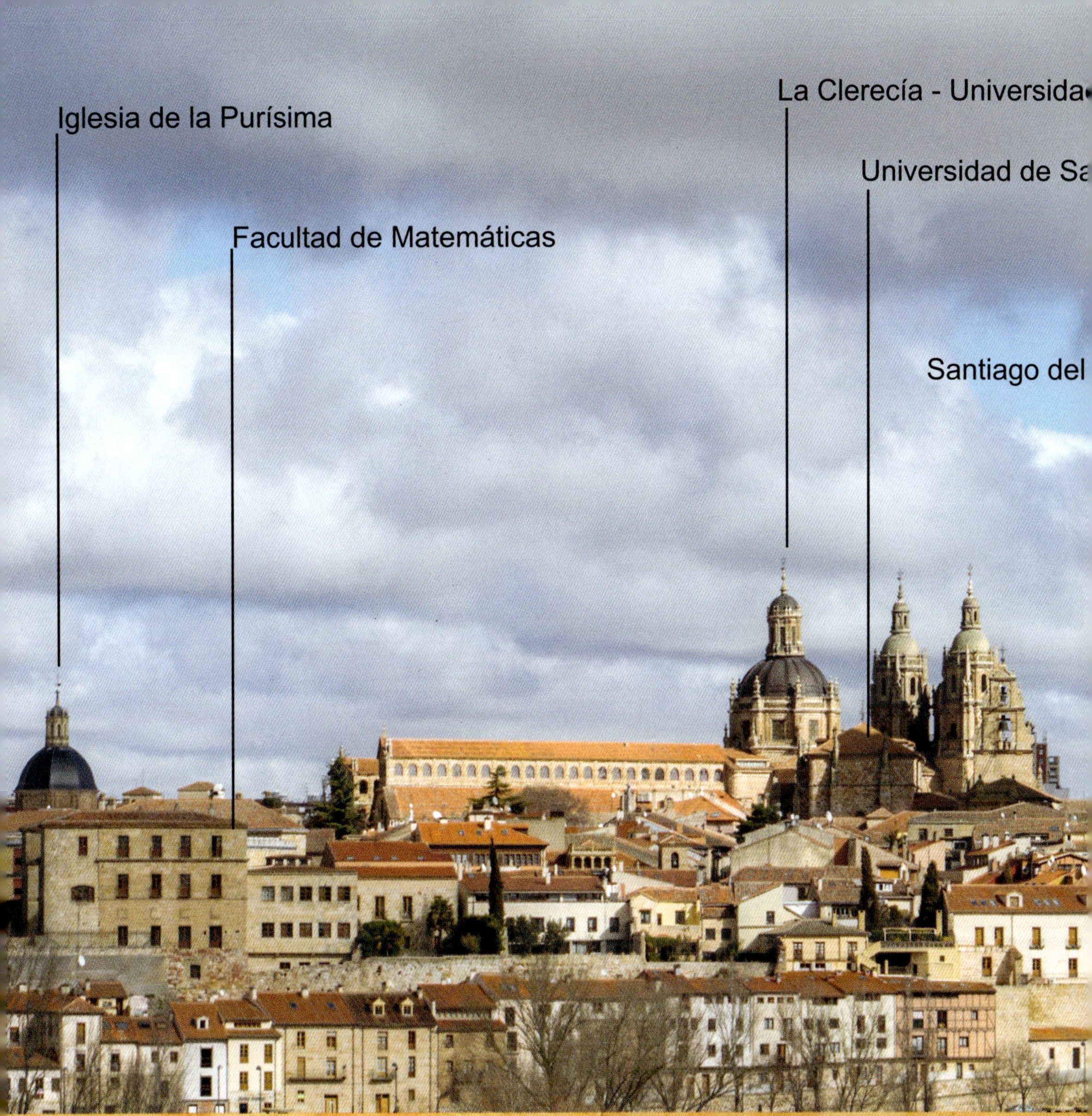
Iglesia de la Purísima
Facultad de Matemáticas
La Clerecía - Universida
Universidad de Sa
Santiago del

SALAM

Pontificia de Salamanca
amanca
Catedral Nueva
Catedral Vieja
Arrabal
Casa Lis

MANCA

TORRE DEL CLAVERO 50

This is the only remnant of the former **15th century** palace which was located here. It is quadrangular in shape and at, halfway up, it comes octagonal, with the lower part as a wall and the decorative show on the upper part made of perfectly cut and squared ashlar; each of its eight sides bears a cylindrical stand supported on corbels with an intertwined decoration and finished in semi-spherical copulins. It is worth mentioning that each of these boxes bears a specific heraldic coat-of-arms, alternatively, of the Anaya family and the Sotomayor family, it being likely that the latter was the craftwork of the building in view of the fact that one of its members, Francisco de Sotomayor, was a beholder of the Alcántara Order, which is where its denomination comes from.

The configuration of this tower recalls the political and social times when it was built, a time at which noble families built urban palaces fortified by higher towers, symbolising their power and political influence; this tower bears witness to this time, with one of them being one of the that are still standing as in the 15th century many were "capped" by order of the Catholic Monarchs, thereby publishing those nobles who had not taken his side in the succession of Enrique IV.

SANCTI SPIRITUS[51]

This church was erected in the **16th century** on the plot of the former Romanesque parish church; in the 13th century it was granted to the nearby St.Anne's convent as a chapel and at the end of the same century it was the responsibility of **Martín Alfonso**, the natural son of Alfonso IX and his wife María Méndez who founded a convent for the Commanders of St.James; nothing is left of said monastery as it was demolished to build dwellings. The design features of the church seem to have provided by **Friar Martin of Santiago**, as it shows stylistic similarities to the church of the St. Stephen's convent[148] where his intervention is documented. It comprises one single nave and a polygonal header.

On the exterior it is worth highlighting the pinnacles and **Plateresque Gateway** where special mention should be made of the large inscription on a slate panel whose content is false and erroneous and was solely carried out to embellish the origins of the royal foundation. Furthermore, it is a specifically Plateresque gateway in which solutions are repeated which have already been seen in other gateways of the city, endowed with careful decoration and medallions with busts which represent, at the spandrels of the arch, the founders and, in the second body, St.Peter and St.Paul. the whole is finished off by a front in whose interior there is a relief carried out by St. James the Moor-slayer. The most appealing thing about the whole is undoubtedly the friezes which develop profane themes of unknown connotations.

GROUND PLAN AND VAULTS

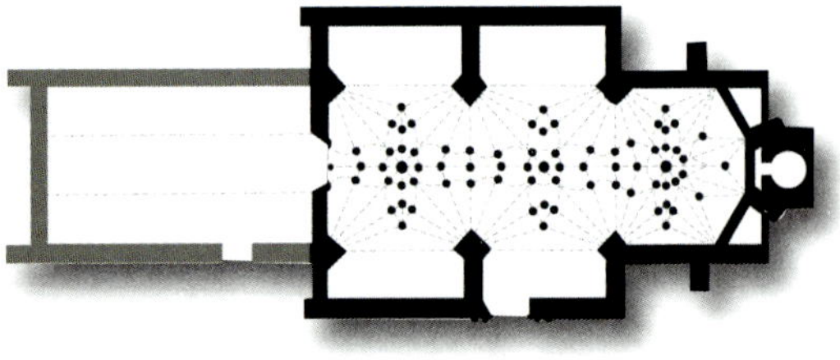

Church, Gothic 16th century

Old choir stall, Renaissance-Mudejar 16th century

In the interior, the main chapel is occupied by an exceptional 17th century **retable** whose sculptures and reliefs were undertaken by **Antonio de Paz**. It is a retable endowed with two bodies and five aisles with a monumental attic. In the first body the reliefs allude to the martyrdom and moving of the body of St.James and on the sides St.Peter and St.Paul. in the second body, in the centre there is a relief of the **Battle of Clavijo** where St.James appears helping the troops of Ramiro I; on the sides, a further two reliefs: the apparition of the Virgin to St.James and the apparition of St.James to Ramiro I before the battle; on the sides, the figures of St.Augustine and St.Ambrose. In the attic the relief is portrayed with the coming of the Holy Spirit or Pentecost. Both the reliefs and the sculptures catch your eye for the broken pleats. The retable is inspired by the one at the cathedral of Plasencia by Gregorio Fernández as well as some of the scenes which derive from the models of said artist. The pedestals depict the evangelists in broken pleated attire and hairstyles deriving from Esteban de Rueda.

On each side of the presbytery lie the **tombs** of the founders, which are severely deteriorated.

At the base of the temple the **choir stall** of the old convent opens out whose **coffered ceiling**, done out in Renaissance-Mudejar style, boasts great quality and mastery. It is here where the image of **Christ of the Miracles** is kept, a crucified piece from the 15th century which is the object of great devotion in the city; it has hairpieces and the skirt and the body are twisted in an S-shape.

REAL CONVENTO DE SANTA CLARA[54]

This convent was founded in 1238 and the church dates from the same century which has undergone many changes and remodelling work over time such as, for example, in the early 15th century as a result of a fire or in the 18th century. The gateway providing access to the church formed by a pointed arch framed by an alfiz belonging to the 13th century. The temple, endowed with just one nave and a polygonal header, is presided over by an **retable** from the early 18th century designed by **Joaquín de Churriguera** and Pedro de Gamboa; but the most important piece is the **frameworked coffered ceiling**, concealed behind the false lunettes vault designed by Churriguera which dates back to the 14th century, highlighting the arrocabe panels with noble coats-of-arms; this space between levels can be visited thanks to a metallic gangway system.
At the base of the church there lies the lower choir stall with the marvellous **murals** from the 13th to the 18th centuries, discovered in 1976 along with

Convent cloister Gateway

Lower choir stall

Simple of a panel with paintings from the lives of the Saints

Coffered ceiling of the cloister

the Renaissance arches and medallions with busts of saints or abbots; the capitals of these arches denote similarities with those of the cloister of the Las Dueñas convent[150]. These paintings correspond to the so-called linear Gothic lineal and they include scenes from the lives of the saints and martyrs as well as from the New Testament with regard to the public life of Christ. After the Discovery and restoration of these paintings, which was worthy of the Europa Nostra prize, the rooms were turned into museums which may be visited today.
During the 18th century the **upper choir stall** lost its original appearance and only a mullioned **Gothic window** remains. In this area there is a small exhibition of popular Nativity cribs of various origins.

The oldest part of the **cloister** retains six old columns from a very early Gothic reminiscent of the Romanesque, deriving from the former St.Mary's hermitage on which the current convent was founded. It is also covered by a polychromated Mudejar alfarje ceiling from the 16th century.
Andrés García de Quiñones also worked on the convent in the mid-18th century and the gateway dates from with time with a stylised niche containing the image of St.Clare.
Furthermore, the interior of the museum houses an ethnological collection including pieces from the early 20th century.

Extrados of the vault and coffered ceiling from the 14th century

SANTO TOMÁS CANTUARIENSE[56]

This temple was founded in **1175** by the English brothers Richard and Randolph which explains the dedication of the temple to **Thomas Beckett**, the archbishop of Canterbury, canonized in 1173. This is one of the few, rare examples of this dedication in Spain, with another church raised under the same denomination in the nearby city of Toro and there is knowledge of another temple, which has now disappeared, in the neighbouring Zamora. We can state that the one in Salamanca is one of the first churches with said dedication in Europe as only five years had passed since the canonization of the saint.

CONSTRUCTION CHRONOLOGY

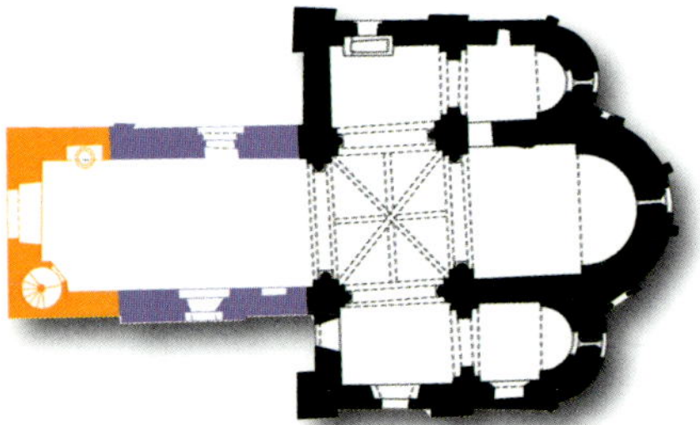

1st stage, Romanesque, 12th century

2nd stage, Romanesque, 13th century

3rd stage, tower, 16th century

The building is in Latin-cross shape, finished off by three semicircular apses; owing to its dimensions, it is thought that in the early days the original project was designed for a basilica round plan with three naves, but the lack of economic resources resulted in the current, much simpler one. Over time it has suffered remodelling work and additions such as the tower and the gateway of the base, built in the 16th century, or the side chapel of the main chapel and plasterwork Baroque vaults, now no longer there, which were added in the 18th century.

The stylistic influences of this temple derive directly from the Old Cathedral[18] where some master or workshop worked who sought to reproduce what had been done there, but demonstrating less skill and command of the material and the resources. Zamoran reminiscences can also be seen in the use of corbels de ovule moulding with coving.

On the exterior the northern gateway is worthy of special mention as well as the header, places where the sculptural decoration is concentrated by means of corbels and chapters; on some closed arches of the header some radial discs appear similar to the Hispano-Roman **funeral steles**, thus serving as an archaizing element.

View of the interior of the church

Discoid stele on the exterior

In the interior is the most intricate, highest quality sculptural decoration, highlighting the transept and the main chapel. Even so, we can distinguish between two different stages or craftsmen in the transept area; the westernmost chapters reveal simple forms with solutions which are more schematic and have less relief, whilst the easternmost ones are worked more intricately and delicate, amongst the latter highlighting the figured capital which represents a combat between two knights, one Christian and the other Moorish. Also present is the sculpture on the brackets which support the transept vault where there is a depiction of masks with large eyes and schematic, very pronounced facial features.

VIRTUAL TOUR GIGA

CONVENTO DE SAN ESTEBAN[60]

HISTORY

The Dominicans reached Salamanca in the first quarter of the **13th century**; in the early days they were established in a convent alongside the fertile plain which was devastated after a flood. The Bishopric thus decided to grant them the Romanesque church of St.Stephen with its land; said church was in poor condition which is why in the late 13th century they erected a conventual unit with a Gothic church on the same plot; at the end of said same century the **General Study for the Dominicans** in Spain was established at this convent.

This convent drove on **Christopher Columbus**' voyage to the Indies, with his main benefactor, and final architect of the voyage, **Friar Diego de Deza**, the teacher of Prince Juan, the son of the Catholic Monarchs. After the Discovery of the New World, it was from this convent in Salamanca that the first Dominicans set sail on an evangelising mission. In this regard its role was very important because, through **Friar Francisco de Vitoria**, the **rights of the natives** were defended.

In the **16th century** it was decided to put up a new convent with a new church; the driving force and stoppage of the work was **Juan Álvarez de Toledo**, the son of the Duke of Alba and the Bishop of Córdoba. The temple we can see today the 16th century one.

During the French invasion the convent was occupied by the Napoleonic troops and used as a barracks; structurally speaking, it did not undergo any major damages; however, by contrast, it was submitted to great plundering. Afterwards, the disentailments and their ruin arrived; the church and the sacristy were saved which ended up fulfilling the parish church purpose.

In the second half of the 19th century, some convent outbuildings were used as a Provincial Museum, thereby dealing with part of its ruin.

In the late 19th century the convent regained its original purpose and became a "General Study" again; after the Civil War it was converted into the Theology faculty.

The Dominican friars were charged with building the bridge which crossed the St- Dominic's stream and the square housing the convent under the guidance of Friar Domingo de Soto; this public work along with the convent constitute a true work of art.

ICONOGRAPHIC PROGRAMME OF THE FAÇADE OF THE CHURCH ST.STEPHEN'S

1 *Adam*
2 *Eve*
3 *Moses*
4 *Elijah*
5 *The Martyrdom of St.Stephen*
6 *Job with his wife and friends*
7 *Solomon*
8 *The Sacrifice of Isaac*
9 *The Calvary*
10 *St.Peter*
11 *St.Paul*
12 *The Eternal Father blessing*
13 *Minerva*
14 *Hercules*
15 *St. James*
16 *King David*
17 *St.Jorge*
18 *Fadrique, 2nd Duke of Alba?*
19 *St. Hyacinth of Poland*
20 *St. Dominic of Guzmán*
21 *St. Francis of Assisi*
22 *St. Catherine of Siena*
23 *Blessed Gil of Santarem*
24 *St. Luis Beltrán*
25 *St Anthony of Florence*
26 *St. Thomas Aquinas*
27 *St. Andrew*
28 *St. Peter of Verona*
29 *St. John the Baptist*
30 *St. Vincent Ferrer*
31 *St. Raymond of Peñafort*
32 *St. Agnes of Montepulciano*
33 *Anthony the Great*
34 *St. Gonçalo de Amarante*
35 *Blessed Peter González*
36 *Blessed Henry Suso*
37 *St. Albertus Magnus*
38 *St. Gregorius Magnus*
39 *St.Peter*
40 *St.Paul*
41 *St. Leo Magnus*
42 *Coat-of-arms of Cardinal Friar Juan Álvarez de Toledo*
43 *Coat-of-arms of the Order of Preachers*

VIRTUAL TOUR GIGA PHOTOS

FAÇADE

The **façade** corresponds to the **retable** type; the whole structure is included within a semipointed arch with caissons which is supported on several buttresses; said buttresses counteract the central nave. It was designed by **Juan de Álava** with the participation of **Friar Martín de Santiago**. The façade-retable is structured through five aisles and three bodies; the three bodies gradually get smaller as they gain height and they are separated by two entablatures. The central aisle is more fully developed; it is coordinated by means of the overlaying of semipointed arches in the first two bodies and a segmental arch in the upper one. This large retable culminates in a Calvary as is frequent in retable art. The main aisle is flanked by two aisles on each side which are coordinated by means of pilasters with italic capitals in which fantasy figures were inserted; niches and canopies structured by means of the overlaying of conic and cylindrical elements were inserted on each of these aisles. All these structures are decorated by scallops and other Plateresque repertoire. The canopies housing the side medallions are in the Gothic tradition; they contain saints of the Dominican order. On the central relief of the second body the stoning of St.Stephen is portrayed, a 17th century work carried out by **Juan Antonio Ceroni**. The decoration of the buttresses repeats the decoration of the side aisles. Many interpretations have been given to this façade: generally speaking, it is thought that it is a large triumphal arch through which three institutions are extolled: the house of Alba, the Dominicans and the Church, which is why on some medallions there are mythical characters and saints on the canopies. It may also have a redeeming meaning; and it may also refer in the first body to the origins, in the second to the Church militant and in the third to the power of the pontificate.

Statue dedicated to Francisco de Vitoria in front of the convent.

PORTICO AND PORTER'S LODGE

Before accessing the interior of the convent it is worth beholding its façade preceded by a **portico**; the latter, through its simplicity, is no less beautiful than the façade of the church with its profuse decoration. It began to be constructed in the late **16th century** in a style similar to **classicism**, but still allowing the coexistence of certain decorative elements whose roots are Plateresque such as the **medallions** with characters of the spandrels or the parapets. These medallions contain portrayals of different saints of the Order; on them there is an entablature of triglyphs and metopes. The portico is finished off by a parapet with Le Escorial style balls, to the liking of their designer **Juan de Ribero Rada**.

On the library façade windows alternate, finished off by a broken triangular front and the coats-of-arms of the Order, except for the central one which corresponds to Friar Pedro de Godoy, the work's patron.

In the interior the **porter's lodge** opens out into a sober space covered by a vault in which the portraits of the most illustrious friars stand out who have passed through this convent: Francisco de Vitoria, Melchor Cano, Diego de Deza, Juan Álvarez de Toledo and Domingo de Soto. The Access door to the cloister is dedicated to the Virgin whose image is situated in a niche on the upper part, flanked by two medallions on which St.Peter and St.Paul are depicted.

Porter´s lodge

CONSTRUCTION CHRONOLOGY AND ROOMS

- *Remains from the 13th and 14th c.*
- *Church started in 1524 by Juan de Álava, continued by Friar Martín de Santiago in 1533, Gil de Hontañón in 1557, Ribero Rada in 1573 and completed in 1610*
- *Cloister, 1528-1544*
- *Soto stairway, 1553-1556*
- *Porter's lodge, Ribero Rada 1591-1592*
- *New chapterhouse and Sacristy, Juan Moreno and Alonso Sardiña, 1627*

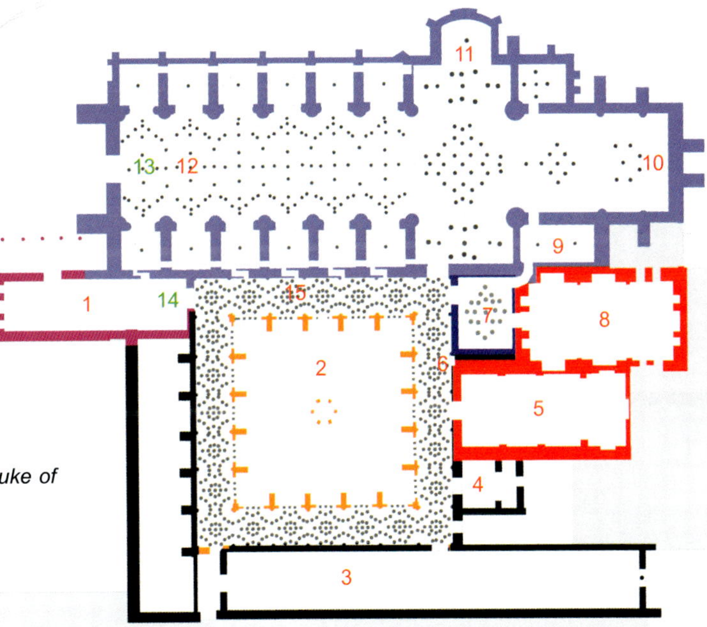

Lower floor

Upper floor

1 *Porter's lodge*
2 *Monarchs' Cloister*
3 *De profundis or Columbus room*
4 *Old chapter*
5 *New chapter*
6 *Old sacristy door*
7 *Soto stairway*
8 *Sacristy*
9 *Bonal chapel, cenotaph of 3rd Duke of Alba*
10 *Main Chapel, main retable*
11 *Rosary Chapel*
12 *Choir loft*
13 *Choir stall, upper floor*
14 *Museum, upper floor*
15 *St. Teresa Confessional*

MONARCHS' CLOISTER

It is believed to be the work of **friar Martín de Santiago** as around about the date of its implementation Juan de Álava was dismissed. It is a cloister of a binary nature: to each lower span there us a corresponding double arch on the upper part. The lower part is heavier because it is coordinated with buttresses which support the tracery vaulting inherited from Juan de Álava and it fits in with a Gothicist approach; whilst the upper part is much lighter and is decorated by purely Renaissance elements, both on the capitals and on the exterior medallions of the spandrels. This cloister provides access to the other outbuildings of the convent. its halls still retain some doors and other remains of the old cloister from the 13th and 14th centuries.

At each of the angles there is a **high relief** with scenes from Jesus' childhood: Annunciation, Birth, Adoration of the Magi and Presentation at the temple; of them all, it is the latter which in the best state of upkeep and it is also the one which has a more intricate carving and great plasticity; worthy of special mention are the treatment of the garments and the representation of the faces, some of which seem true portraits that lend greater realism to the scene.

Between the arches of the cloister there is a series of medallions with the busts of prophets from the Old Testament, identified by name. This type of decoration fits in with the Plateresque style. At each of the corners of the cloister a door opens, four in total, highlighting the access door to the Soto stairway whose design could have been carried out by Rodrigo Gil de Hontañón, but the master builder was someone else. It has a semipointed arch flanked by pilasters on which an entablature rests ad non which there are vegetal forms which are finished in torches and a oculus with the figure of St. Dominic in the centre.

Soto's staircase from a landing.

SOTO STAIRWAY

It was built in 1539 and it takes its name from its patron **friar Domingo de Soto** one of the most famed theologians from this convent who took part in the Council of Trent.

It is a cloister type stairway endowed with four flights in each flight is supported in the preceding one, affording the sensation of lightness; it has been designed by way of a cimborrio. The vault is characteristic of the work of **Rodrigo Gil de Hontañón**, carried out in line with transepts, sagging and tierceron which are finished off in crow's feet; it all goes to form a kind of flower which frames the keystone. There are corbels at each of the corners which are enclosed in an impost running around the whole of the upper part of the quadrangular space. In the centre of each side a semipointed span with a Renaissance profile opens out and flanking it there are pairs of medallions with the Church Fathers and the four evangelists.

On the side opening out onto the cloister there is a three-arched stand with a balustrade at whose spandrels cherubins have been placed, repeating the type at the tower of Monterrey palace[96]. In the stairway recess there is decoration with caissons with rosettes and on the last flight there is a relief of Mary Magdalene recumbent, perfectly adapted to the space.

SACRISTY

The **sacristy** is a great classicist work from the first third of the **17th century** sponsored by **Friar Pedro de Herrera** and undertaken by **Alonso de Sardiña** and **Juan Moreno**. It is a rectangular building which is divided up into three sections by some high pilasters corintias on which a frieze of triglyphs and metopes runs where, on occasions, figurative reliefs alternate. The space is covered by a semipointed vault with lunettes. Each of the sections is decorated by an architectonic structure comprising an arch flanked by Doric pilasters on which a frieze runs and a broken triangular front is raised, finished off by pyramids and El Escorial style balls and the coat-of-arms of the patron.

Both at the base and at the header the wall is endowed with a structure whose configuration is reminiscent of a triumphal arch with niches on which sculptures are situated carried out by **Antonio de Paz** which represent, at the base, St.Vincent Ferrer and St. Anthony of Florence and the upper one St. Dominic of Guzmán; at the header are St.Peter and St.Paul, flanking the Assumption of the Virgin. On the section immediately after the header is the burial place of the patron, represented by a sculpture also carried out by Antonio de Paz.

View of the new chapter

OLD AND NEW CHAPTERS

At one of the halls of the cloister the Old Chapter opens out, also called the **Pantheon of Theologians** as the remains of the most illustrious Dominican theologians are buried here: Francisco de Vitoria, Domingo de Soto and Matías de Paz, inter alia. This room was also used as a meeting place where important decisions were made, not only for monastic life, but for life outside too, as it was from here that Columbus' trip to America was urged as well as the sending of missionaries to new territories. This room was in use until the first third of the 17th century when the New Chapter was built at the behest of Friar Iñigo de Brizuela; its architects were Alonso de Sardiña and Juan Moreno which carried it out in line with the classicist style which prevailed at the time. The room looks austere and sober, almost totally lacking in any ornamental elements, reduced to those characteristic of the architecture as it was going to be adorned in paintings and tapestries from the private collection of the works patron.

View of the Sacristy

THE CHURCH

Access to the church is gained via a beautiful Plateresque door carried out by Friar Martín de Santiago, highlighting the niche with the image of St. Joseph with the Baby Jesus, a work by **Antonio de Paz**. Typologically speaking, the convent church follows the model of a **preaching church**. It is endowed with a sole nave with chapels between the buttresses and a transept marked heightwise, but not in terms of width; the chapels are joined by shortcuts which allow access to them without interrupting worshipping. The header is highly developed as it was initially thought of installing the choir stall here. Despite the fact that the construction of the church lasted for many years, unity was achieved as the different architects who worked there were governed by the design features initially provided by **Juan de Álava**. Despite following the 15th century models, transparency was achieved by the proportion between the height and the width as well as by the absence of pillars and the unity of the vaulting system. The vault type is probably of the most complicated ones which have been designed by Juan de Álava; a circular warp is used around the keystone, drawing around it six ogee arches of differing widths. At the side chapels the drawings are different but vaults endowed with Gothic tracery are still used. The keystones of the vaults of the whole church are decorated by means of medallions on which there is a depiction of the coat-of-arms of the Álvarez de Toledo as the promoters of the works, but the coat-of-arms of the Order and Dominican saints also appear. Juan de Álava was followed at the head of the works by **Friar Martín de Santiago** who closed the nave and the chapels and endowed them with retables, also starting the transept and the presbytery, but the latter were completed by **Rodrigo Gil de Hontañón** who adapted the project to Renaissance tastes. The cimborrio is surrounded by a platform on which the body of lights is raised with three spans flanked by medallions. The ribs no longer reach the ground, but rather they are supported on corbels which rest on italic capital columns.

The retables of the transept are dedicated to the Virgin of the Rosary, St.Thomas Aquinas and St. Dominic of Guzmán. The first was commissioned to José de Churriguera whilst the other two are the work of **Joaquín Churriguera**, both being endowed with canvases painted by **Antonio Palomino**.

The main retable was carried out between 1691 and 1693 by **José Benito Churriguera** having been commission by Friar Pedro Matilla, the prior of the convent and confessor of Carlos II. It is a retable endowed with large dimensions, with the giant column on a monumental scale predominating; it boasts the dynamism characteristic of the Baroque and served as the model for the other Baroque retables of the so-called churrigueresque type. However, the decoration is still contained and does not fill it all, creating this sensation of horror vacui which is so characteristic of the work of the Churriguera. The stipe is introduced and this is one of the first times it appears. The theme around which the iconography of the retable is developed is the **Eucharist**, with the monumental tabernacle - with a shrine-type structure – thus serving as the centre of the composition. The motifs appearing on the tabernacle include a canopy with cloth glued run by angles, a very characteristic Churriguera theme. At the base of the tabernacle there are two angels in very graceful, dynamic postures. The sculpture is limited to the sides with two images of St.Francis and St. Dominic of Guzmán. The Solomonic columns are covered by vegetal decoration based on bunches of grapes and vine leaves. In the attic we can find a huge canvas by **Claudio Coello** with a depiction of the **martyrdom of St.Stephen**, reinforcing the Eucharistic meaning of Christ. On the sides two angels appear again on volutes.

THE UPPER CHOIR STALL

This is accessed via a beautiful day with fine decoration presided over by the coat-of-arms of the cardinal friar Juan Álvarez de Toledo. This space comprises a large surface area occupied by walnut stalls from the **17th century** comprising 118 ceremonial chairs designed by **Alonso de Balbás**; their decoration is very sober and contained as is characteristic of Classicism. However, the most prominent element of the choir stall is the large **mural** situated at the base; it was carried out in fresco by **Antonio Palomino** in the early 18th century. Its title "*The Church Militant and Triumphant*" is highly revealing as it is telling us about the content thereof and its composition is inspired by a work by **Rubens** endowed with a similar name: *The Triumph of the Church*. It is signed by the author on the wheels of the cart. This painting is characterised by its refined composition in which the cleanliness of the drawing and the chromatic variety stand out; it is rife with symbolism and boasts a complex iconographic programme which is successful in terms of order and accuracy.

CONVENT OF THE DOMINICAN DUEÑAS[61]

This convent was founded in **1419** at the houses granted to this end by Mrs.Juana Rodríguez, the wife of Juan Sánchez de Sevilla, Royal Tax Collector of Castile, with King Juan II. There are still important remains of the original palace such as some **Mudejar arches** providing access to various rooms which retain their forms and decoration based on different-coloured glazed tiles; this original construction also includes a beautiful, simple coffered ceiling which is kept today at the Fine Arts Museum[48] of the city, assembled in one of its rooms.

Thanks to new donations, the mid-**16th century** saw the construction of the **church** and the **cloister**, thereby providing order and access to the different buildings which went to make up the monastery which is why the cloister is endowed with an irregular pentagonal shape on its sides of different lengths.

On the exterior the monastery denotes a sober, simple appearance, concealing the sculptural wealth we find within its cloister; only the **gateway of the church** is noteworthy whose authorship is not documented, but it could well be attributed to **Juan de Álava** or the Dominican **Friar Martín de Santiago** owing to the similarity with other works carried out by these masters. On the main body the well-made image of the Virgin with the Baby Jesus is worthy of special mention and it still retains some of the polychromy in some areas of the niche; flanking it there are two medallions at the spandrels of the arch which represent St. Peter and St. Paul. However, without a shadow of a doubt the "must" element in the visit to this monastery is its **cloister**, a Renaissance masterpiece. It is organised at two heights; the lower part consists of arches lowered on columns at whose spandrels there are **medallions** with depictions male and female characters who are difficult to identify. The upper gallery is architraved with twice as many holes as the lower one; the lintels rest on richly adorned **blocks** with figures endowed with impossible expressions and postures and imbued with a fantastic, mythological nature. Another sculptural element which is worthy of special mention in this upper gallery are the **medallions** running along

Church gateway

Upper cloister

like a frieze both on the exterior and on the interior of the entablature. Various motifs are represented here: coats-of-arms, flowers and busts, with the latter being the most interesting; they show a variety of characters such as, for example, saints and martyrs with their identification attributes such as St. Catherine of Siena, characters from the time or from the Bible; this is how St. Peter is portrayed with the keys, King David with the harp, Veronica with the holy handkerchief, Moses with the tablets or the Virgin with the Baby Jesus. Some of these medallions are matched, forming scenes such as the **Annunciation**. Here we can observe that they were undertaken by different people, but they do show a common thread which is their Roman-like character in terms of the conception of the characters, their attire, movements and gestures as well as their morphology which are reminiscent of reliefs and sculptures from Classical Antiquity such as, for example, the medallion of the Virgin with the Baby Jesus which is clearly inspired by Roman matrons.

And you can't leave the monastery without trying the exquisite **pastries** made by the Dominican Mother superior in their bakery.

The Annunciation: matching medallions, the Archangel Gabriel and the Virgin receiving the Holy Spirit

Cloister of the Dominican Mothers Superior convent

PLAZA DE COLÓN[65]

Statue erected in honor of Christopher Columbus on the IV Centenary of the Discovery of America; the work of the great sculptor Eduardo Barrón (1858, Moraleja del Vino, Zamora - 1911, Madrid)

PALACIO DE ORELLANA[65]

The building's construction was ordered by Alonso de Anaya y Barrientos, the canon of Toledo and chairman of the Chancellery, in the late **16th century;** it is believed that Juan de Ribero Rada worked on this building. Its façade is very sober and anything decorative is subject to architectonic, an interest is only retained in the ornament and the rhythmic treatment of some elements such as the alternance of triangular and bent frontons or the use of Ionic and Tuscan Pilasters. This constitutes a change in artistic thinking, a change in artistic conception, already close to the purism and sobriety of Classicism.

TORRE DE ABRANTES[66]

Only the tower remains of this **15th century** medieval palace; it should be higher. The original element comprises the gateway and a mullioned window on the side façade; this is a good example of Hispano-Flemish architecture.

OTHER MONUMENTS OF INTEREST

CASA DE LA CONCORDIA72

Also known as the House of the Battles, of the original building from the **15th century** only its gateway and famous inscription remain: *Anger generates hate, harmony nourishes love.*

Its current name derives from the tradition whereby this was the place where **St.John of Sahagún** pacified the two noble factions who were confronting each other over power in the city in **1479**.

SAN POLO74

Built in the first quarter of the **13th century** in **Mudejar Romanesque** style. At present only the stumps of the walls and the southern and western elevations.

EL CARMEN DE ABAJO75

Belonging to the St. Andrew's Convent of the Barefoot Carmelites, its foundation dates back to the late **15th century**. Only the chapel remains, undertaken by **Manuel de Larra Churriguera** in the mid-**18th century.**

SANTIAGO DEL ARRABAL2

The church we can see today is a **reconstruction** of the one which was here in the 12th century. It is known that the original temple had three apses.

HOLY TRINITY - ST. PAUL62

Only the façade remains of the convent building of the Barefoot Trinitarians convent, following the architectonic model created by Francisco de Mora.

ST.KATHERINE'S SCHOOL59

This was wounded in the late **16th century** as a School. Only the gateway remains, comprising two pilasters supporting a broken front in whose centre the coat-of-arms of the School is located.

ST.ILDEFONSO SCHOOL57

This was founded in the early **17th century** as a School. The façade is only adorned by the coat-of-arms of the founder and a Latin inscription above the door.

CALATRAVA SCHOOL58

The Military Order of Calatrava decided to found a school for its members in the City. The new building which started to be built in **1717** under the supervision of **Joaquín de Churriguera**; they were resumed under the orders of **Jerónimo García de Quiñones** and with the collaboration of **Gaspar de Jovellanos.**

THE HOLY TRINITY GENERAL HOSPITAL[55]

It was founded in the early **13th century,** but it was until the late 16th century when it was converted into the most important hospital in the city, combining around twenty small hospitals. In the late 18th century it was endowed with an anatomic amphitheatre designed by John of Sagarvinaga. From its first age the 16th century **Gothic chapel** has been retained, carried out by **Pedro de Ibarra**; it is here that the remains of the founder of the order are kept, **Mother Bonifacia Rodríguez de Castro**.

SAN CRISTÓBAL[52]

This Romanesque temple was founded in the mid-**12th century** by the Knights Hospitallers of St. John of Jerusalem. In the exterior it has a notable collection of figurative **corbels**.

HERMITAGE OF OUR LADY OF MERCY[53]

It was founded in **1389** as a hostel for pilgrims. It has two gateways; one of them is in Plateresque style, whilst the other one is in Baroque style and is attributed to **Andrés García de Quiñones**.

SAN JUAN DE SAHAGÚN[35]

This church is dedicated to the patron saint of the city. This building is a clear example of the **revival** movement; built between 1891 and 1896, it bears Neo-Romanesque and Neo-Gothic elements. Its façade contains two bronze reliefs which are the work of Aniceto Marinas which narrate two episodes in the life of St. John of Sahagún which occurred in Salamanca: the miracle of the Yellow Well and the "Pacificación de los Bandos".

HOUSE OF THE SEXMEROS DE LA TIERRA[49]

Its name derives from the four councillors of the Town hall, representatives of the agricultural boroughs of the City Council who kept the ownership of the building in the 18th century, but the building dates further back to the late **15th century.**

ST.JULIAN AND ST.BASILISA CHURCH[48]

The origin of the building is **Romanesque,** worthy of special mention is the **gateway** and a high-relief of a fantasy animal similar to a lion. Attention should also be paid to the **tomb** of Francisco Rodríguez del Manzano, the first Count of Francos and the teacher of Carlos II, carried out by **Juan de Setién Güemes.**

CORPUS CHRISTI CONVENT$_{33}$

It was founded in the mid-**16th century** as the convent of St. Clare Franciscans. The Renaissance church gateway stands out, denoting stylistic similarities to the gateway of the Solís House[95], owned by the promoters of the convent.

SAN JUAN BAUTISTA DE BARBALOS$_{31}$

It was founded in **1139** by the **Knights Hospitaller.**

It bore witness to the sermons of **St. Vincent Ferrer** in the city in the 14th century.

OUR LADY OF MOUNT CARMEL$_{32}$

Today, we can only see the church of the Barefoot Carmelites' Convent; it began to be built in the early 16th century following the features of **Fray Jerónimo de la Madre de Dios.**

COLEGIO DE SAN AMBROSIO$_{3}$

It was founded to give shelter to needy children; in around **1720** it suffered a renovation by **Joaquín de Churriguera.** In the façade appears the image of St.Joseph with the Baby Jesus, both sculptures having been carried out by **José de Larra Churriguera**.

HOUSE OF JUAN DE OVALLE OR SANTA TERESA$_{30}$

It was the house chosen by **St. Teresa de Ávila** for her first foundation in the city; the Saint from Ávila provided a description of this house in chapter XVIII of the *Book of Foundations*, expressing her wariness as it had been a student house; this was the place where she suffered the **ecstasy** of Resurrection Sunday, after which she wrote her famed: *I live without living in myself, and in such a way do I hope, that I die because I do not die.* **Mother Bonifacia Rodríguez de Castro**, the founders of the **Servants of St. Joseph** alongside **Father Butiñá** in 1874, arrived at the house in 1881.

BIBLIOTECA DE NUESTRA SEÑORA DE LOS ÁNGELES$_{6}$

It was founded in the mid-**16th century** by Jerónimo de Arce and later, at the end of the 18th century, it merged with the former San Millán School. After falling into disuse it was converted into a university library. Of the original building the gateway is retained where it is worth mentioning the image of the Virgin framed by an alfiz and flanked by the statues of St.Peter's and St.Paul. It bears the characteristics specific to the age with Gothicist motifs and Renaissance elements.

HOUSE OF THE ALDERMAN OVALLE PRIETO OR DE UNAMUNO$_{28}$

This was the home and place of decease of **Miguel de Unamuno**; this house belonged to the city alderman Juan Antonio Ovalle Prieto; it is a building from the **18th century** in which its gateway stands out, a bulging relief.

ST. MARY'S OF THE KNIGHTS CHURCH$_{27}$

This church was founded in the 12th century; it was highly transformed in the **16th century**. In the 18th century the side chapel was built.
The famous architect **Juan de Álava** or Ibarra and part of his family were buried at this church.

PALACIO E IGLESIA DE SAN BOAL$_{36}$

The palace was built in the final decades of the **15th century** and all that remains from this time are the gateway. In the 16th century it is worth mentioning in its interior the double-height Renaissance **courtyard**.
The original church was carried out in the 12th century under the guidance of St. Boal; in the mid **18th century** it was rebuilt. The only notable thing is the sculpture of St. Boal, to be found in a niche on the door which is attributed to **Simón Gavilán Tomé**.

MONTELLANO PALACE$_{42}$

Although its current name derives from the dukes who bought the building after the Disentailment, in reality its promoter was Alonso de Paz in the late **15th century**. Over time it was sold to the Barefoot Trinitarians who turned it into a convent.

CHURCH OF THE TRINITARIANS$_{43}$

Built in the **17th century**, back-to-back with the Montellano palace, all that remains is the façade.

RODRÍGUEZ DEL MANZANO PALACE$_{45}$

This late **15th century**/early 16th century building.

FIGUEROA PALACE$_{44}$

Built in the mid-**16th century** by order of Juan Rodríguez de Figueroa, an advisor to Felipe II and a professor at the University; the architect who provided the design features was **Rodrigo Gil de Hontañón**. In the late 19th century it was reconverted into a Casino where the famous literary circle used to meet up which was attended by Miguel de Unamuno.

PLAZA DE LOS BANDOS

The Bandos square is situated on the former plot of the Romanesque church of St. Thomas which brought together one of the noble factions in conflict during the 15th century and whose war ended with the intervention of St. John of Sahagún.

HOUSE OF DOÑA MARÍA LA BRAVA[39]

This building from the late **15th century** receives said name in reference to one of its previous residents, **María de Monroy**. The house we can see today, it belongs to the late 15th century and it was erected by her grandson Gonzalo Rodríguez de Monroy on the plot of the former palace; it preserves the original façade, it has a broken alfiz which incorporates the door and the upper balcony.

A TALE OF REVENGE

Mrs. María de Monroy, the widow of Enrique Enríquez de Sevilla with whom she had two sons and one daughter, received her nickname from the revenge wreaked with the murder of her two sons at the hands of the Manzano brothers. The conflict between the youths has to be set within the framework of the war between the two rival factions which at that time were fighting for power in Salamanca, the St. Thomas faction, to which María belonged, and the St.Benedict faction. The confrontation between the youths arose owing to a ball game that culminated in the death of the Enríquez; when their mother found out, she pursued the murderers as far as Portugal where they had taken refuge and killed them, bringing the decapitated heads of the Manzano to Salamanca and depositing them in the tomb of her children in the parish church of St. Thomas.

ST. ELIAS CONVENT OF BAREFOOT CARMELITES[40]

Popularly known as El Carmen de Arriba, it is the only one preserved of the old college. It is now a parish church.
The temple was erected between 1694 and 1703, following the features of Friar Antonio de Jesús María and the ground plan of the St. Teresa of Ávila church.

SOLÍS PALACE[41]

As regards the original **15th century** building where Felipe II and María of Portugal were married in 1543, little remains, just a **window** on the side of the Neo-Plateresque tower, said window bears the coat-of-arms of the Solís held up by two wild people.

GARCI GRANDE PALACE[38]

the latter belonged to the lords of Zaratán and it was built in the mid **16th century**, denoting stylistic similarities with the Rodríguez Figueroa Palace[158], leading us to think in the same author: **Gil de Hontañón**. Two aspects stand out in the original building; firstly, the medallions with human, on the other hand, the two chamfered windows.

First published 1999

ISBN 0 9537414 0 0

Published by Michael Dane, Walton-on-Thames, Surrey. Tel: 01932 221271

Printed by Ian Allan Printing Ltd, Riverdene Business Park, Molesey Road, Hersham, Surrey KT12 4RG

Front cover: **Watercolour painting of Ashley Park House from Ashley Road by unknown artist, c.1890.** *(Elmbridge Museum)*

Selected Bibliography

Special acknowledgement is due to Stanley Jackson's *The Sassoons, Portrait of a Dynasty* (London: Heineman 1968), Cecil Roth's *The Sassoon Dynasty* (London: Robert Hale 1941) and Michael Blackman's *History of Ashley Park, Walton on Thames* (Walton and Weybridge Local History Society 1976)

Other sources consulted or quoted are listed below:

Bolitho, Hector *King Edward VIII* (London: Eyre and Spottiswoode 1937)
Falk, Bernard *He laughed in Fleet Street*
Haweis Mrs., *Beautiful Houses* (London: Sampson Low 1882)
Hyamson, Albert *The Sephardim of England* (London: Methuen 1951)
Jenkins, Roy *Gladstone* (London: Macmillan 1995)
Magnus, Philip *King Edward the Seventh* (London: John Murray 1964)
Sassoon, Seigfried
Diaries 1923-1925 (London: Faber 1935)
The Old Century and Seven More Years (London: Faber 1938)
Complete Memoirs of George Sherston (London: Faber 1972)
Thornycroft, Elfrida *Bronze and Steel* (Shipston on Stour: King's Stone Press 1932)
Wilson, Jean Moorcroft *Siegfried Sassoon, The Making of a War Poet* (London: Duckworth 1998)
The Pearl of Days. An intimate Memoir of the Sunday Times 1822-1972.

Contents

	Selected Bibliography	2
	Map of Ashley Park Estate	4
	Foreword	6
I	The family background	9
II	The early years of David Sassoon & Sons	12
III	Sassoon David Sassoon (S.D.) – The early years	15
IV	The first Sassoon in England	18
V	Flora as Guardian	29
VI	Flora – The later years	36
VII	Alfred and his family	40
VIII	Rachel	49
IX	Joseph, Louise and their young family	57
X	Education of the children	68
XI	Sport at Ashley Park	73
XII	The children fly the nest	81
XIII	The Boys go to War	87
XIV	The Home Front 1914-1918	93
XV	Running down	99
XVI	What became of the survivors	105
	Family Tree	109
	Index	110

Ashley Park showing the boundaries at the time of the acquisition by the Sassoon family in 1863.

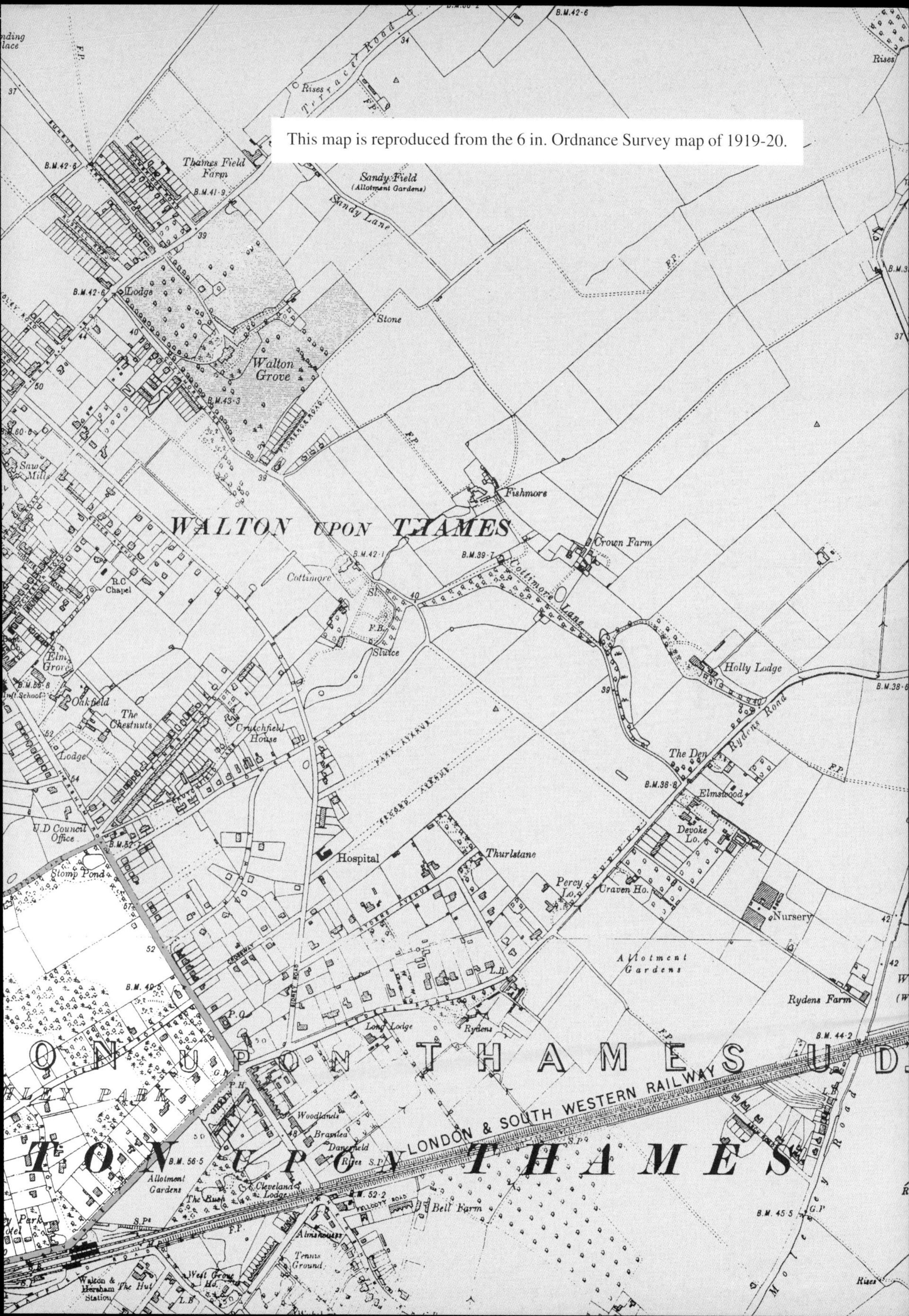

This map is reproduced from the 6 in. Ordnance Survey map of 1919-20.

Foreword

In the year 1968, I purchased a house in Ashley Park Avenue in Walton on Thames and in a desultory manner sought to find out something of the history of the Ashley Park Estate on which my property had been built in the1930s. From the then Weybridge Museum (now the Elmbridge Museum), I was able to establish that the Fletcher and Sassoon families had owned the Mansion and the Park from the late 18th century until shortly after the Great War, when the Estate was sold and broken up for development.

This scant information was supplemented in 1976 when the Walton and Weybridge Local History Society published a booklet entitled 'The History of Ashley Park, Walton on Thames' by Michael E. Blackman. This booklet concentrated on the historical background of the House and its owners rather than on the lives of its occupants, but it had the added advantage of pointing me in the direction of two further books on the Sassoon family, who had been the last family to own the Estate and occupy the mansion as a family home.

The first of these was 'The Sassoons-Portrait of a Dynasty', by Stanley Jackson which had been published initially in 1968 and which was subsequently published in a paperback edition in 1989. This book gave me a much greater insight into the branch of that family who had resided at Ashley Park, but its main emphasis was on the family businesses and on those members of the family who were involved therein. As the only Ashley Park Sassoon to be involved in the family business was Sassoon David Sassoon who died in 1867, less than four years after he had purchased the Estate, only a small part of the book related to the affairs of Ashley Park. It however provided a most valuable source of the family background in Baghdad and in Bombay, having been written with the help of the family then living.

The second book 'The Sassoon Dynasty' by Cecil Roth, published during the Second World War proved more difficult, being out of print, but I was able to borrow a copy through the good offices of the Surrey County Council library service. This book not having been written with the support and assistance of the family, added comparatively little information to that of the Jackson book, but it did provide some pointers as to the whereabouts of the descendants of Ashley Park Sassoons.

There my interest and knowledge of the Sassoons appeared to have come to an end. However at the beginning of 1990 my business career in the City of London dictated that I move offices to 85 Gracechurch Street. Much to my surprise I discovered that from offices in this building the E. D. Sassoon Banking Company Limited had managed its European operations from the early 1930s until the hub of the business moved to the Bahamas in 1950, following the communist take over of China which made the continuation of the business in Shaghai impossible.

This coincidence of my second occupation of premises which had formerly been closely connected with the Sassoon family suggested to me that I should carry out further research into the Sassoons of Ashley Park. The opportunity so to do would have to await my retirement, which was not to occur until late in 1994. Since that date I have carried out a considerable amount of detailed research, which at the time was done solely for my own interest. In the course of that research, I met with or spoke to as many descendants of the Ashley Park Sassoons as I could trace, who were to prove most helpful. In the course of my discussions with Joseph Sassoon, the grandson of the Joseph Sassoon who owned Ashley Park House from his coming of age in 1876 until his death in 1918, he expressed the view that it would be unfortunate if all my research was not to be published, a proposition which I had not considered, nevertheless I undertook to explore this further, hence this product.

I am therefore particularly indebted to Joseph Sassoon who provided me with a wealth of family detail, including a little gem of his father's memories of Ashley Park which he wrote down in old age. Additionally he gave me access to family photographs, many of which appear in this book. He also enabled me to meet his brother, Jacques and other members of the family who provided additional insight into life in Ashley Park in their parents' time. Without such insights this narrative would have lacked a good deal of its colour. In addition members of the Sassoon family have contributed financially towards the costs of publication, for which I am most grateful.

I have also received much help and encouragement from Neil White, the Manager of Elmbridge Museum who took a great interest in my researches and convinced me that the book was worthy of publication. The museum has also agreed to assist financially towards its publication, as has Phoenix Property and Finance Ltd.

In addition The Charity of Robert Phillips has made a generous contribution.

I must also acknowledge the assistance of the many who provided research facilities or background information including:-

The staff of The Imperial War Museum.

The newspaper archives of The Surrey Herald, The Surrey Comet, The Sunday Times and The Observer.

The archivists of Sandroyd School, Eton College, Christ Church and Exeter Colleges, Oxford and the Royal Military College, Sandhurst who provided details of the education of the growing Sassoon family.

Mr Harry Steer provided me with information on his extensive golfing family from their early beginnings in Ashley Park.

To all of them and to others unnamed with whom I have talked from time to time, I extend my grateful thanks.

SASSOON OF ASHLEY PARK

Chapter I

The family background

A family of Jews bearing the name Ibn Shoshan were living in Toledo, Spain in the twelfth century and their descendants continued to live there until they, together with all the other Jews living in Spain, were expelled from the country in1492. They were members of the Sephardic branch of Judaism which had flourished there for many centuries. Most of these refugees from Spain set up home in the Netherlands with others moving to the other countries of Western Europe including England. A minority however gravitated eastwards and linked with the well established Sephardic Jewish communities of the Middle East, and in particular Mesopotamia.

The Sassoons with whom we are concerned could first be identified with certainty living in Baghdad in the eighteenth century where they were part of a significant Jewish minority in that predominately Islamic city. The link between the Ibn Shoshans of Toledo and the Sassoons of Baghdad, claimed by leaders of the family in later generations after they had flourished in England, is difficult to find convincing. The main branch of the Sassoon family also claimed to trace their descent by the direct line from King David, as did many other prominent Jewish families.

The Jewish community thrived in Baghdad and, as in other communities, grew to fill the role of bankers. In the course of time a Chief Banker was appointed from the Jewish community in Baghdad, who also became the treasurer for the city, in which capacity he collected taxes from his fellow Jews as well as acting as their spokesman. The role was, of necessity, filled by one of the wealthiest Jews in the city, who was greatly respected and known as Nasi or ‘Prince of the Captivity’. Although several members of the Sassoon family were understood to have held this office which was of a quasi hereditary nature, the first who is capable of verification was Sason ben Saleh, born in 1750, who became Nasi in 1778 at the early age of twenty eight and enjoyed the honorary title of Sheikh. By this time Baghdad was entering a period of decline which led to the status of the minorities becoming increasingly precarious. Sheikh Sason having married, produced seven sons the eldest of whom died in infancy, thus his heir was his second son David, born in 1792, who was the first to adopt the present spelling of the family name.

As was to become the custom, David started in the family business almost as soon as he could read and was quick to show an aptitude for the business. His work typically occupied almost all his waking hours. As was the practice amongst the Jewish families in Baghdad at that time, his father was not slow to arrange his betrothal when he was only fifteen years old, choosing Hannah Joseph, the daughter of a leading merchant, who was two years younger. It was not until more than ten years later that their first son was born shortly after the Sheikh had retired as Nasi in 1817, worn out by the growing intrigues as successive governors came and went. Suleiman Pasha with whom the Sheikh had established a good working relationship had by that time been forced to flee the city and been replaced by Daud Pasha who dashed David's expectations by appointing Ezra ben Rahel as Nasi. This appointment did not last but David now aware of the precarious situation wisely declined Daud's subsequent offer of the succession. This enraged Daud who became his sworn enemy but such was David's status that he remained unofficial leader of, and spokesman for, the increasingly oppressed Jewish minority. By 1826 he had taken over most of his father's public offices and had fathered two sons and two daughters by his first wife, Hannah who had since died. He had recently remarried Farha Hyeem, the daughter of a prominent Jewish Baghdad merchant. Daud sought a trumped up charge on which to have him arrested and imprisoned in order to use him as a bargaining tool to expropriate the Sassoon family fortune by ransom. The Sheikh duly paid what would have been the first of many such extortions to obtain David's release. He had had the foresight to charter a ship to take David to Basrah which departed the same evening, from whence he immediately proceeded onward to Bushire in Persia which was out of reach of Daud's pursuing forces.

The unhappy David, now aged thirty six, was therefore forced to live apart from Farha, his second wife who was not yet out of her teens, his four children by his first wife and his aged father, an exile in a strange country. He had however two significant assets, firstly a small amount of the family funds that he had been able to bring out of Baghdad and secondly the goodwill and the contacts his family had made with the traders of the Middle East over a number of years. In addition his father was still active in Baghdad in various financial transactions amongst the trading community there. With his modest capital it was prudent for David to act as a middleman rather than to take the greater risks as an importer in his own right and he quickly built a reputation for honouring his commitments. As well as trading with Baghdad and Mesopotamia he extended his contacts to India, acting through the East India Company there. Meanwhile the Sheikh was making plans to bring the remainder of the family out of Baghdad which was to cost him a significant part of his fortune. This he achieved in the early part of 1830 but at a considerable cost to his health. Shortly after being reunited with his son, David, but before his eldest grandson, Abdullah would celebrate his Barmitzvah, he died in Bushire. Thus the family had severed their immediate connection with Baghdad. However both David's third son Sassoon David and his great grandson,

Arthur (of Ashley Park) were, to return to Baghdad, Sassoon David to study, and Arthur as an officer in the victorious British army of 1917.

Bushire was never going to be able to meet the aspirations of the ambitious David, whose interest was drawn to Bombay, where the trading was controlled by the East India Company, under the benevolent rule of the British, who permitted all races to live and trade together and allowed freedom of religion. He carried out a reconnaissance which confirmed his earlier impressions as to the attractions of setting up in business there. He was able whilst there to elicit promises of support from the British Agent and the Jewish business community, many of whom were emigrants from Baghdad. On his return to Bushire he found his wife to be expecting their first child and his mind was finally made up when a family friend who had provided him with a warehouse in Bushire agreed to finance a ship load of goods to stock his first godown in Bombay. This assistance enabled him to leave Bushire for Bombay where thus stocked, he was able to set up business at 9 Tamarind Lane in the early days of 1832 and embark on a business career which was to succeed beyond even his wildest dreams. Later in the same year a son, Sassoon David Sassoon was born who was, nearly thirty years later, to be the first of the Ashley Park Sassoons. This use of the family name followed by the father's forename was not uncommon in Jewish families of that period as it was felt that it * would confuse the Angel of Death who called once a year. This style was to be repeated by Sassoon David's first born son Joseph, in the naming in turn of his eldest son. To avoid confusion and to distinguish him from his father, bearing in mind the similarity of their names, he will henceforth be referred to as S.D.

* the author's confused. Jews avoid naming a son the same as the father for the reason given. The <u>middle</u> name does not ~~the~~ represent the son's own name but is for "ben = son of", understood.

Chapter II

The early years of David Sassoon & Sons

David's timing in the setting up of his business in Bombay could not have been better as it coincided with relaxation of the East India Company's monopoly of trade between London, India and the Far East. This new freedom to trade had the effect of sparking off a major commercial expansion. He immediately reopened his former links with Baghdad and correctly foresaw that the growing trade with England would lead to a demand for sea front wharves in Bombay. He therefore used a substantial part of the trading profits earned in the early years for the acquisition of such properties. As trade grew David found that the goodwill associated with the Sassoon name which had been built up over a long period in Baghdad opened up wider business opportunities such as the acceptance of Bills of Exchange. Despite their youth the two sons of his first marriage, Abdullah and Elias, followed their father's example and took an active role in the firm's activities under the watchful eyes of their father who continued to be the decision maker. By this time David had become widely recognised as the leader of the Arab/Jewish trading community in Bombay. The family continued to grow in number as further sons arrived, Reuben in 1835, Arthur in 1840 and the twins Aaron and Solomon a year later, all of whom would in due course share in the management of the business. There were also three daughters, who in line with the practice of that time, were not considered for management positions in the firm. A further son, Frederick born in 1853, would complete the family. They all were encouraged in their faith, became fluent in both Hebrew and English, as their father eagerly embraced the freedom provided under British rule. As the business grew so did the profits but David was rigid in enforcing the rule that only the direct family were to manage the business, thus placing a great strain on himself and his two eldest sons. Of the two, Abdullah was the more outgoing and made friends easily whereas Elias was withdrawn and resembled his father in appearance, but they had both been brought up to follow David's main business dictum, the avoidance of impulsive decisions. By the time of Queen Victoria's accession to the throne in 1837 the firm was prospering and reinvesting the profits, both in property and in trading in an ever increasing range of merchandise. Larger premises were acquired to meet the requirements of a growing staff of clerks, mostly recruited from the expanding Jewish community, who were finding the freedom of the Indian regime more attractive than life in Baghdad in a minority religion. Their faith was encouraged, a room in the firm's premises having already been set aside for their daily prayers.

As Abdullah and Elias showed their ability to handle the growing trade, David felt able to divorce himself from the every-day matters and to apply his energies in the search for wider horizons for business expansion. He had for some time taken an interest in the affairs of the Bombay Cotton Exchange and had formed the view that this was a business with opportunities, but with his usual caution he was not to be rushed. India had been a producer of cotton for over a hundred years but it was not renowned for quality, largely due to inadequate storage facilities. For this reason the operators of the spinning looms of Lancashire had a preference for the more expensive American cotton. Two events convinced David that the time was right for a move into this trade. The failure of the American crop in one year created a panic demand from Lancashire for Indian cotton with prices forced to unprecedented heights. Although this level of demand was short lived the continuing high prices of American cotton made Indian cotton an increasingly attractive alternative. Secondly, in 1840 the Bank of Bombay commenced trading mainly to provide the financing for this growing trade. David made his move, having seen that the scope for export of the raw material could be matched by the import of Lancashire manufactured goods for onward trade in the markets of the Middle East. His early ventures into this market were typically at a modest level but were later to have great impact on both the profits of the firm and on the life of his sons, in particular S.D. When the first mills opened in Bombay, David was able to greatly increase his export of cotton and import of the resulting finished goods for sale in his existing markets in Persia and Mesopotamia.

He also recognised China as a major growth market for the future but as yet he felt that his capital base was insufficient to justify the risk of challenging the might of the established firms like Jardine Matheson, so he bided his time. The two way trade generally between India and China was very lucrative, being largely based on the import of tea and silks balanced by the export of opium and to a lesser extent, cotton. The trade in opium was giving rise to increasing friction between the Chinese authorities and the traders, mainly the East India Company, who had the support of the British government. This provoked what became known as the First Opium War in which the gunboats of the East India Company together with their troops were used with the tacit approval of the authorities in London. The weak Chinese resistance was no match for these forces and the war was soon over. The resulting Treaty of Nanking in 1842 opened up Shanghai and four other ports outside the legal jurisdiction of the Chinese Empire to operate with tax free concessions. By 1844 David felt strong enough to test this market, sending his second son Elias to China, where he quickly assessed that Shanghai and Hong

Kong offered the best prospects for trading in textiles and opium as well as the financing of shipments and providing small scale merchant banking activities for other traders. Returning to Bombay, he left a deputy in Shanghai, but by 1850 it had been decided that the rapidly expanding trade justified his full time presence there and he accordingly opened the firm's headquarters on the Bund. With his prudent ways learned at his father's feet, Elias avoided the reliance on the quick profits from opium and built a diversified trading operation with probably the best wharves in the Far East. By 1858, following the Second Opium War, David Sassoon & Sons had established not only a firm footing along the whole Chinese coast but had expanded in a small way into Japan as well.

David had remained the sole owner of the firm until 1852 when Abdullah was admitted into partnership and was joined shortly thereafter by Elias following his success in Shanghai. The sons by David's second marriage would not be taken into the partnership until after his death. The firm's growth was so dramatic that by 1857, at the time of the Indian Mutiny, it had become one of the most powerful in the East. This success was achieved by the vision of David and by his hard work and that of his sons who had been well trained from an early age. There were at this time great opportunities for success but many firms became over stretched and failed, but David, although cautious, grasped those opportunities and continued to invest the profits both in the expansion of the business and in real estate, in particular in wharfage. It was also said that one of the main reasons for the firm's success was the use which David made of his eight sons.

David had been brought up by his father to give support to charitable organisations both in the Jewish community and in a broader context. The firm's growth in profits enabled him to give generous support to these charities. He financed, amongst other institutions the Poona General Hospital and the Sassoon Mechanics Institute which were for the benefit of all sects. Despite many offers the only public appointment which he accepted was that of Justice of the Peace, a role to which he applied himself with great diligence. In 1853 he had been one of the first to swear allegiance to the Crown, becoming naturalised and a great anglophile despite the fact that he still spoke little English.

Following the Mutiny the East India Company was forced to cede its political powers to the Crown. David marked this occasion by hosting a banquet for five hundred of the leading citizens of Bombay at his mansion which included as guest of honour the new Governor, Lord Elphinstone. In his speech proposing the health of David Sassoon, the Governor said "We must not forget that at the time of the Mutiny, when threatened with danger and whilst some were panic-stricken, Mr Sassoon and his family were the first to come forward in support of the British Government."

Chapter III

Sassoon David Sassoon (S.D.) – The early years

S.D., the eldest of David's sons by his second marriage grew to be tall and thin, but from an early age had not enjoyed the robust health of his elder half brothers, nor was his manner as easy going as that of his younger brothers. He gave the appearance of being a dreamer, although like all his brothers he applied himself diligently to the business from an early age. He spoke with a slight lisp and did not give the impression of gaining enjoyment out of life. He was twelve years younger than Elias and only three years older than Reuben, the extrovert of the family, with whom there was a gap in temperament. Accordingly he was, and remained something of an odd man out in a family much given to internal feuds.

He was apprenticed to the counting house but unlike his elder brothers he had outside interests and made the time to write extensively for, and to edit the first Hebrew-Arab periodical published in Bombay, a journalistic trait to be followed by his, as yet unborn, daughter some forty years later. He was well suited to edit such a publication, being fluent in Hebrew, Arabic, Persian and Hindustani as well as English. He also showed great interest in the Jewish religion which his father recognised and encouraged by sending him to Baghdad for study and initiation there into biblical lore. Thus some twenty years after David's flight his third son returned there at the age of eighteen. During his time there a marriage was arranged for him with a girl four years younger than himself. She was Farha Reuben, the daughter of a wealthy and devout merchant, well known to David. So father and son now had wives of the same forename, to become a source of some confusion in the years to come. Dark and small in stature, as her husband was tall, she was strong in character, quick tempered with an unpredictable temperament which fluctuated between extreme kindness and haughtiness as the mood took her. S.D. and many others over the years were to be dominated by her.

Returning to Bombay, their first son Joseph was born on the last day of 1855, by which time Elias had returned from his initial posting in Shanghai. Abdullah relieved him there as a short term measure while the family agonised over the succession. Despite S.D.'s frail health and inexperience in business it was decided to send him to Shanghai as Elias' replacement notwithstanding the recent birth of his son. As was the practice in the family Farha did not accompany her husband to China but went to live with her father in law in some style in his imposing mansion on the Malabar Hills overlooking the Bay of Bombay. In Shanghai he showed that

S.D. at the time of his arrival in England.

his training had fitted him well to manage the business there in a period of growth, but the humid climate caused a worsening of his somewhat fragile health.

By 1857 the fast growing cotton trade with England, coupled with evidence of a more liberal attitude to Jews there, convinced David of the desirability of opening a branch office in London and the question of its staffing became a matter for decision. The candidates for the post were S.D., who had shown a safe pair of hands in Shanghai but who suffered breathing difficulties in the oppressive heat of China which was not helped by his smoking, a well entrenched family habit. He had also been the first of his family to take to western dress which was considered an asset for business in London. The main opposition came from his younger brother Abraham, who had by now adopted the name Arthur, who was keen for the position. Reuben supported Arthur's claims but after a family discussion involving David and his elder sons Abdullah and Elias, it was decided in favour of S.D., who was willing to take up the challenge. His claims were pressed behind the scenes in Bombay by his wife from her position of strength in David's household and who was by now expecting her second child. He was considered to be the right choice as Arthur, not yet eighteen, was felt somewhat immature and unproven in independant management for a role which David felt would have a significant impact on the future prosperity of the firm. Arthur blamed Farha for his rejection which gave rise to a rift between the two branches of the family which was to last for fifty years and was only healed in the final year of Arthur's life.

Accordingly, following the birth of his daughter who was named Rachel, S.D. set sail for England in the latter months of 1858, leaving Farha and his two young children to join him as soon as he was established there. His arrival coincided with the Jews achieving greater acceptance in England by being admitted to the House of Commons by swearing a separate oath of allegiance. Leopold de Rothschild was the first of his race to take his place in the House of Commons. It was not long before the Sassoon family would become widely known as the Rothschilds of the East and were to cement this title by inter marriage with the Rothschilds in England.

Chapter IV

The first Sassoon in England

S.D. arrived in London in the winter of 1858 and acquired the crown lease of 17 Cumberland Terrace as his first home in London. Cumberland Terrace, on the eastern edge of Regents Park was the most splendid of the terraces surrounding the Park. Like its neighbouring terraces it had been designed by John Nash, completed some thirty years earlier and named after Ernest, Duke of Cumberland, one of George IV's younger brothers. Nearly eighty years later this house was to attain a certain notoriety when Edward, Prince of Wales (briefly to become King Edward VIII) took a lease on it for his wife to be, Wallis Simpson, in the difficult period prior to his abdication.

S.D. also negotiated the lease of the firm's first London office at 8 Leadenhall Street in the City, appropriately almost next door to East India House where he regularly lunched and met with fellow merchants trading with India. East India House had been built in 1800 and was quite the most imposing building in Leadenhall Street. As well as offices it contained a grand concert hall, many paintings and architecture of the sub-continent, noble statues, a library of Indian literature and a museum containing oriental exhibits. Following the demise of the East India Company the building was no longer appropriate and was demolished in 1862, the contents being transferred to the Victoria & Albert Museum which had opened five years earlier. The demolished building was replaced in 1866 by East India Buildings, a much less impressive office block, into which the Sassoon Company offices later moved.

He now eagerly awaited the arrival of Farha, soon to be anglicised to Flora, and their young children, Joseph and Rachel in the spring of 1860. With the family now reunited they began to settle into a routine in their adopted country. S.D. worked hard at the office, initially without the pressures of his previous appointment in China. He began to collect Hebrew manuscripts and prepared translations for learned Indian periodicals. The main place of worship for Sephardic Jews in England was the Spanish & Portuguese Synagogue in Bevis Marks which was built in 1701, and which was conveniently situated only a few hundred yards from the Sassoon office in Leadenhall Street. It was natural that S.D. should worship there and he quickly made his mark, being elected a Yahid (ordinary member) in 1861 and at once throwing himself into the charitable work of the Jewish community. Flora settled into London along with Joseph aged three and baby Rachel.

Whilst the proximity of Regents Park provided some open space close by their house, it was clearly not the ideal place for a growing family in their early years.

In addition the cold and the London fogs which they experienced in those early winters together in England convinced them of the desirability of finding a home with more space outside the capital, particularly as Flora had by now given birth to a further son, who they named Alfred, and was expecting her fourth child. Other businessmen with whom S.D. was now mixing freely had country estates and he resolved to join them without moving too far from London whilst retaining Cumberland Terrace as a London residence close to his place of business. His third son, named Frederick, was born in the summer of 1862 and before the end of the year they were actively looking for an estate appropriate to their requirements. As S.D. was about to become a landed gentleman he approached the College of Heralds with a view to obtaining a coat of arms. This was duly granted in December of 1862 and gave a fresh impetus to the search for their new home. The shield which was designed by the College in discussion with S.D. reflected his roots, including a palm tree on a gold background between a pomegranate and an olive branch. Above this was a lion (of Judah) on a blue background. The crest above the shield showed a dove (from the ark) carrying an olive branch. The motto was surprisingly in french as 'Candide et Constanter', which could be translated as 'sincere and determined'.

Early in the following year S.D. was put in touch with the Fletcher family, which had long established connections with the East India Company over several generations and who were seeking to find a buyer for their Ashley Park estate in Walton on Thames, some eighteen miles south west of London. It would appear that the purchase was arranged privately between the parties who would have been known to one another through their common interests at East India House. It was necessary for David to agree to the financing of the purchase of this estate, for S.D. was not at that time a partner in the firm and his capital was limited although he personally owned property in Shanghai, the personal ownership of property being a practice which David had encouraged his sons to follow. David quickly agreed to finance the purchase, which was finalised in the October of that year at a cost of £48,500 (equivalent to £2m at today's prices). The 1861 census had shown that by that date the Fletcher family had vacated the main house, which was let, but Sir Henry Fletcher, the fourth baronet was still in residence in Ashley Cottage, formerly the Ranger's House on the estate.

The land had formed a part of the hunting chase of Hampton Court Palace, created by Henry VIII in the late 1530's, and was situated close to the Palace of Oatlands. These lands remained in Royal hands until they were dechased early in the reign of Edward VI. The original building accounts, which are still extant, show that an existing house was pulled down and a much larger house constructed of locally made bricks was built between 1602 and 1605 at a cost of £2,610 for

Lady Berkeley, the wife of the 7th Lord Berkeley, who held a Crown lease. The estate then passed through a number of hands until it was acquired in 1718 by Richard Boyle, the 1st Viscount Shannon. He employed Sir Edward Pearce who had designed Parliament House in Dublin to carry out significant alterations to the house adding an additional wing, remodelling the jacobean windows and converting the long gallery into a ballroom with Corinthian columns. Richard Boyle had fought as a volunteer at the Battle of the Boyne in 1690 and enjoyed a successful military career rising to be the Commander in Chief in Ireland in 1720 with the rank of Field Marshal. For nineteen years he was a Member of Parliament and was appointed one of the Lords Justices in Ireland two years before his death in 1727.

In her will Lady Shannon, who had continued to live in Ashley House until her death in 1755 left £1,000 to be spent on a monument to be erected in their memory in St. Mary's parish church in Walton. Her daughter, Lady Middlesex, a friend of the Prince of Wales in the reign of King George II who was a regular visitor to the house, employed the sculptor Louis Roubilliac to create the monument, which has recently undergone repair and which is considered one of his finest works.

The Estate continued in his family and in 1787 had passed through the female line to Sir Henry Fletcher the first baronet, of Clea Hall, Cumberland. He was

Ashley Park House in the Regency Period. *(Elmbridge Museum)*

Ashley Park House, front view. *(Elmbridge Museum)*

already the Member of Parliament for Cumberland and continued to be so until his retirement in 1802. Shortly thereafter he was able to more than double the acreage of the Estate by taking advantage of the Walton on Thames Enclosure Award of 1804. His son, the second baronet, also Henry inherited the property after his father's death in 1807 and pulled down the wing added less than one hundred years earlier by Lord Shannon. The second baronet made little other impact either nationally or locally and died in 1821.

In turn his son, the third baronet also named Henry succeeded to the estate and took a greater interest in local matters, including the Parish Church. Between 1835 and 1851 he fathered no less than twelve children, all baptised in the parish church. When he in turn died in 1851 the fourth baronet, confusingly also named Henry, was only fifteen years of age and his uncle John, the only surviving brother of the third baronet acted as his guardian and trustee during his minority. The fourth baronet came of age in 1856 and after a short spell in the army, married three years later without showing any interest in Ashley Park, probably because of the high cost of its upkeep. His uncle, John who had moved to Derby Lodge, across the river in Sunbury appears to have negotiated the sale to S D., which was speedily agreed by the Fletcher family trustees.

This then was broadly the estate acquired by S.D., comprising the main red brick Tudor mansion, built in the shape of a H, somewhat modified and said to be wide enough for five carriages to be lined up across its front, set off by splendid magnolia trees with ivy growing up the front. Another house of some size on the borders of the estate named Ashley Cottage, originally the dower house, then the Ranger's Cottage but more recently occupied by Sir Henry Fletcher. In all the estate purchased was approximately 430 acres, slightly less than at its peak, as some minor sales of peripheral land had taken place.

An article recording a visit to house in 1880 gives an account of the rooms which had been little changed since the Fletcher era. The main entrance was on the east front opening into a spacious hall rising to the full height of the building, with a gallery at the southern end. The hall was panelled in oak, surprisingly painted over in imitation oak graining by the Fletchers in the late eighteenth century. The panels contained an undistinguished portrait of Henry VIII, a number of recent portraits and old weapons including halberds and crossbows. From the hall rose the main staircase, of old oak with a renascence style balustrade painted with gilt scrolls of agapanthus, supported by carvings of young boys. On the ground floor

Ornamental staircase.

to the right ran a long corridor leading to a sitting room painted in white and gold with a mantelpiece designed like a broken pediment of a classic temple. The panelled dining room was most attractive with portraits of Viscount Shannon and his family, who were mainly responsible for the interior design of the house. The walls of the drawing room were covered in wallpaper depicting large branches with birds of paradise in strong colours. On the left from the hall was a pretty octagonal room on whose corners and shelves were displayed oriental china, one of the few rooms modified by the Sassoons. The eighteenth century marquetry and cabinets displayed treasures in amber and embroidery worked in pearls.

Beyond this room was the ballroom, or banqueting hall, the most attractive room in the house running the full length of the south wing, with many windows. the ceiling was supported by a group of Corinthian pillars with tall, arched doors between them. Above the hall a long gallery connected the two wings of the house. Upstairs were a number of panelled bedrooms, each with old oak double doors and old hinges. The only room of merit on this floor was a spacious billiard room appropriately furnished. Other items calling for favourable comment included several old ceilings, a few pieces of 17th century carving, an old press and mantelpiece in oak which reached the ceiling. Each with a fireplace which at best was capable of providing only inadequate heating. There were servants' quarters appropriate to such a house at that time.

As befitted its considerable age the house was extremely difficult to heat and costly to maintain and would be a source of constant expense in the future. Candle lighting was still the vogue as the supply of gas was not to be connected for a further six years. Mains water was also a resource of the future, the house then being supplied from a well at the south west edge of the estate from which it was pumped into a large storage tank mounted above the nearby ice-house conveniently situated at the highest point of the estate. The house was supplied from there entirely by gravity.

The land to the north fell away beyond the Italian garden, whose feature was an old stone fountain, and the terrace towards the River Thames. The river at Walton was first bridged in 1750, but replaced by a more substantial bridge in 1786. The two main arches of this bridge had collapsed in 1859 so that at the time of S.D.'s purchase Walton was temporarily isolated from the County of Middlesex. This bridge was the subject of a painting executed in 1802-4 by J.M.W. Turner which can be seen in the Victoria Art Gallery in Melbourne, Australia. The replacement bridge of latticed wrought iron girder construction was not to be erected until 1864 when it opened as a toll bridge. It was not until 1870 that passage over the river was free of charge. Walton then had a population of some 2,000 persons, mostly employed in agriculture and domestic service.

Italian garden, Ashley Park.

The park contained many attractive trees and was highly regarded for its stately cedars of Lebanon, as well as specimen oaks, elms, limes, planes and scotch pines. The Long Drive which led to the house from the gates in Walton High Street was over half a mile in length and was flanked by oak trees, many of which survive to this day. There was also the Short Drive which led from what is now Ashley Road which bisected the estate. The Long Drive was rarely used by the by the Fletcher family, a practice followed by the Sassoons other than when visiting the village. Accordingly the gates leading onto the High Street were normally kept closed. The two drives converged before passing through two pillars, still extant, to enter the garden area with its wide lawn fronting the house. The equally large west lawn was divided from the park on which deer and cattle roamed by a ha ha, parts of which can still be seen in private gardens.

S.D. proudly mounted his recently acquired coat of arms above the front door when they moved in, where it remained for over sixty years latterly suffering from the ravages of the weather. Together with Flora he set about furnishing the mansion in style with Chippendale furniture from their adopted country. The Chinese room was hung with ornate silk and contained lacquered cabinets from the Far East. Ottomans and carpets reflected their background in the Middle East and India, while imposing paintings hung on the walls. As a student of Jewish history, S.D.

fitted out the library to contain his growing collection rare Hebrew manuscripts and books, with sufficient space for additions which were steadily being acquired. The oak-lined hall had yellow glazed windows rising to the full height of the building which gave the room a golden glow in the sunshine. The growing prosperity of the firm meant that the fitting out of the property could be carried out with little or no regard to cost. Many of the required servants were inherited with the property, supplemented by additional recruitment needed to look after the family together with an army of labourers and gardeners to maintain the estate. Flora ruled over them in an autocratic manner, having become used to dealing with a retinue of servants during the years she spent in David's house in India, when S.D. was in China, and during his early years in England, before she joined him. Something of a bully, she quickly established her authority but doubtless her oriental manner took some getting used to by the local staff. Thus S.D. had arrived as the first of his family, but by no means the last, to be appropriately housed as an English country gentleman. The estate factor, Elwood was soon installed in the recently vacated Ashley Cottage, where he continued to reside until after S.D.'s death.

Although the American Civil War had started in 1861 its impact on the business in London was not immediate but it soon became clear that the storm clouds were gathering. The growing two way trade in cotton between Bombay and Lancashire quickly made it important for the firm to open branch offices in Manchester and Liverpool in order to handle the business locally. The increasing efficacy of the Union blockade of the Confederate ports was beginning to make the export of cotton from America, on which the economy of the southern states largely depended, increasingly difficult. Consequently the problem of obtaining the adequate supplies of raw cotton so badly needed to keep the Lancashire mills running increased both the demand for supplies from India, and the prices obtained for them. The Sassoon Company was uniquely able to satisfy this need from its operations in Bombay, taking advantage of their extensive wharfage and godowns together with their offices in London and Lancashire. Accordingly the family firm prospered as never before and S.D. found less time for his books, but at the same time was able to show his metal as a businessman. By 1863 the blockade of the southern American ports had become highly effective and in Lancashire there was panic that with the supplies of cotton drying up, there would be widespread unemployment there. To avoid this catastrophe supplies were urgently needed from elsewhere as prices were driven ever higher. The Sassoons both in London and in Bombay had foreseen this and were well prepared to meet the demand and prospered as prices and profits soared to levels not previously achieved. They were not alone in making these high profits but unlike many others they foresaw that there would come a time when prices would fall and with their customary caution they were not to be exposed when that time came. The London office thus became the hub of the activity in cotton and S.D. was not found wanting as he travelled to and from Lancashire to supervise the branch offices and to meet local merchants. The train network which had been developed over the previous thirty years had reduced the journey time from London to Manchester to five hours but nevertheless all this travelling took its toll on S.D.'s never robust health, and caused Flora growing concern. His breathing problems, fuelled by his heavy smoking, had now been diagnosed as emphysema, which with the added pressures of business, was now showing signs of strain on his heart. He by now had also acquired a number of directorships in several important businesses unconnected with the family business as his mercantile reputation grew in London.

In his spare time he was able to follow outside activities and became both a Fellow of the Society of Arts (FSA) and a Fellow of the Royal Society (FRS). Taking a lead from his father he gave generously of both his time and his money to charities, mainly, but not exclusively, with Jewish connections. Education interested him and he became a member of the Council of Jews College for the training of Rabbis and served on the committee and was examiner in hebrew for the Jews Free School. Among other donations he gave £1,000 towards Gladstone's work amongst the poor in the notorious Seven Dials area of London. Although not politically active he generally supported the Liberal party which formed the government for almost all of his time in England. William Gladstone came from a Liverpool family and as his Hawarden estate was not far from there in Flintshire. He and S.D. would probably have first met through S.D.'s business connections in Lancashire. In 1865 Gladstone, previously a Member of Parliament for Oxford University became Member for South Lancashire providing further opportunities for them to meet on S.D.'s visits to that part of the country. The Sassoons certainly became friendly with both William and Catherine Gladstone, which friendships were carried into the next generation when despite their many pressures William and Catherine found time in a busy life to attend the wedding of S.D.'s daughter Rachel.

S.D.'s elder brother Abdullah, now anglicised to Albert, paid a visit to London in 1864 and was impressed by the way in which the English operation was being run. Everything seemed to be set fair in the Sassoon empire but in November of that year whilst in London, Albert received the news that David, the patriarch of the family, had died peacefully at his home in Poona. He was buried in the grounds of his synagogue there according to the customs of the Jews of Baghdad. The widely spread network of sons attended services in Bombay, Hong Kong, London and Shanghai. The Governor of Bengal, Sir Bartle Frere suggested that David be commemorated by the erection of a statue of him to stand in the Sassoon Mechanics Institute in Bombay, an organisation for which David had provided the finance for its building some years earlier. The Governor headed the subscription list which included from England, Gladstone, the Rothschilds, Mocattas and Montagus and more surprisingly the Lancashire mill workers. The statue was commissioned from Thomas Woolner R.A. and work thereon commenced. David's estate was variously estimated as between £2 million and £5 million but the higher of these figures would appear to be nearer the mark. After charitable bequests and provision for his unmarried daughter the eight sons shared the residue equally so it is reasonable to assume that S.D.'s share was in excess of a quarter of a million pounds, (over £10 million at today's values). The heavy upkeep of the Ashley Park estate was therefore taken care of for the foreseeable future as S.D. also enjoyed the income from his own investments and from his share of the growing partnership profits.

Shortly afterwards the American Civil War came to an end and as the Sassoons had expected, the price of Indian cotton collapsed as better quality American cotton once again became available. Panic followed in Bombay, with the failure of the Bank of Bombay adding to the woes as many of the firms trading in cotton went under. However the Sassoons avoided this catastrophe as it was their policy not to hold large stocks and accordingly they survived the war even more powerful, but without their leader of the last thirty years. Following David's death Albert and Elias initially continued to work together and in consultation with S.D. decided to use their position in the Manchester Cotton Exchange to buy piece goods in bulk while the prices were low for resale in India and the Far East. S.D. continued to work hard and must take a share of the credit for successfully piloting the English offices through the vicissitudes of the market in cotton, a success not generally acknowledged in the family. In addition he had become increasingly involved in the management of the Bevis Marks synagogue and by 1866 he had become a member of the Mahamad (Warden). The synagogue finances which had been precarious for some time had taken a turn for the better helped by S.D.'s management and generosity. In furtherance of his faith he planned to follow his father's lead by building a synagogue in the grounds of Ashley Park, but due to his untimely death this project was not carried out although he still envisaged embarking on it up to the time of his death. However, the unforeseen pattern of events of the last five years had prevented the managership of the London office from being the expected sinecure and the strain had a severe adverse effect on his health, which had never been robust, much to Flora's concern.

In India, Elias who had been happy to play a secondary role in the firm while his father was alive, did not relish the continuance of this position now that Albert was the chairman of the firm. By 1867 it was clear that a break was imminent as Elias was planning to go on his own. This inability of members of the family to get on with one another was to be a feature of many branches of the family through three generations, much to their cost. Elias duly opened his new business as E D Sassoon & Co which flourished initially in a similar business to that of the senior firm. In later years it also branched out as a merchant bank and in due course was to out perform the original Sassoon Company, both in India and in London. In the same year Flora was presented at Court by Lady Mary Wood, Alexandra, Princess of Wales receiving her.

As Thomas Woolner's statue of David neared completion in London it was arranged that it would be exhibited in the Victoria & Albert Museum in South Kensington before shipment to India. S.D. had been invited to a preview of the statue at the museum on the of 23rd June. In the midst of a heat wave he had travelled up from Walton on the previous evening to spend the night in his house in Cumberland Terrace. On the following morning he took a cab to lunch at the Langham Hotel in Portland Place which had been opened by the Prince of Wales two years earlier. After lunch, whilst waiting there for a carriage to take him on to the museum he collapsed and died, aged thirty five as the strain on his heart proved too much. Thus, at the early age of thirty one, Flora was left with four children between eleven and four years of age but without any of her relatives living in her adopted country.

S.D. was buried as befitted a pillar of the Jewish faith in the new Jewish cemetery in Mile End, East London, where he was later to be joined by his widow, his three sons, a daughter in law and one granddaughter. The Rev. Dr. Artom in his oration made reference to the great loss to his community, his great benevolence and his kindness of heart. He referred to a life which showed promise of great usefulness, cut short in its prime. His death came a year too soon for him to see his friend William Gladstone become Prime Minister in the first election following the increase in the electorate as a result of the Reform Bill.

It was now clear that the successful piloting of the firm through the period of enormous growth arising from the American Civil War and its aftermath, together with his other activities, took a heavier toll on his health than his brothers realised from their occasional visits to London. Neither did he receive from them the credit due to him for avoiding the many pitfalls which brought down other firms during this period. Flora could however see the damage being done to him and never forgave the members of the family firm for his untimely death. Unfortunately it turned her against the firm in particular and business in general to the extent that no member of her family of the succeeding two generations either joined the family firm or made a worthwhile career in commerce. She was nevertheless not too proud to accept a pension from the firm which enabled her to live out her life in comfort. Reuben was chosen to succeed his brother as head of the London office and shortly after arriving in London followed him in the synagogue in Bevis Marks.

Chapter V

Flora as Guardian

The provisions of S.D.'s will are of considerable interest as they were to be followed to the letter by Flora and thus played an important part in shaping her thinking with regard to the bringing up of her four children. In addition they were to become a source of friction and legal dispute involving his three sons once they had each attained their majorities. The will was made in August 1865, not long after his father's death, clearly took into account his share of David's estate. The trustees appointed under the will and who would act during the minority of the four children were his wife Flora, who was also appointed sole guardian of the children during their respective minorities, his brothers Elias and Solomon and his confidential clerk Edward Perugini. As Elias had by the time of S.D.'s death already left the family business in less than harmonious circumstances and Solomon unlike his other brothers was never to take up residence in England, Flora had unfettered control of the upbringing of the children as Edward Perugini had also disappeared from the family scene shortly after S.D.'s death. He resigned as a trustee within four years, being replaced by Antonio Sefi, a friend of the family and a successful City merchant with an Indian background. Flora was also enjoined to ensure that the children adhered faithfully to the Holy Religion of their forefathers and that they inter marry amongst their own religion. In the case of Rachel, in addition she should not marry without the consent of her mother. Flora was to reside at Ashley Park during Joseph's minority unless she should remarry and she was to enjoy the net rents from any lettings from the estate. S.D. left an estate in England valued at over £2m (over £70m at today's prices) as well as properties in Bombay and Shanghai. To Flora he left an annuity of £1,000 per annum for life, Alfred and Frederick were each to receive a lump sum of £5,000 and Rachel £10,000 in trust with the income therefrom for life. He also provided for the completion and maintenance of the proposed synagogue in Ashley Park should the project have been commenced during his lifetime. As the work had not been started by the time of his death this bequest lapsed. After charitable gifts in Baghdad, Bombay and London together with other small bequests the balance of his estate including the Ashley Park Estate was left to Joseph on attaining his majority. Twenty years later the bequest leaving the entire Ashley Park Estate to Joseph was to be successfully challenged in the Courts by Alfred and Frederick, burdening the estate with substantial legal costs. Joseph also inherited S.D.'s collection of Hebrew books and manuscripts on which he was to build as it became one of his major interests for the remainder of his life.

Reuben, an altogether more outgoing personality, succeeded S.D. in charge of the London office and moved to England, setting up house in Lancaster Gate, where he was to enjoy a busy social life and to grow increasingly close to the Prince of Wales and the Marlborough House set. He was soon to be joined in London by another brother, Arthur as the hub of the family firm began to move from Bombay to London. David's widow Farha also came to England where she took up residence in Brighton. Reuben found time amongst his other interests to take an interest in the religious upbringing of the Ashley Park children, satisfying himself that they were learning the Torah. However he and S.D.'s widow, Flora, were always at odds with one another over the details of the children's upbringing. Joseph celebrated his Barmitzvah without problems while Rachel, Alfred and Frederick continued to receive instruction in Hebrew from a visiting rabbi. Reuben rightly expressed his concern that Flora, in the absence of a male influence, was spoiling the children but he made no constructive suggestions for remedying the situation. It is doubtful whether she would have allowed Reuben more influence over the children's upbringing as she, not without reason, considered him something of a playboy.

Flora took her role as chatelaine seriously, but unsurprisingly took no part in the life of the village, other than to continue to permit the Park to be used by the villagers for annual fetes, flower shows and other significant events. Influenced by her conviction that business pressures had been the cause of her husband's death she used her influence steadily to build up in the minds of her children a distaste for business. Thus this part of the family which turned out to be the most prolific branch of the Sassoon dynasty was not able to provide the additional management that the firm sorely needed in later years when such leadership was not available elsewhere from the next generation of the family. Thus with no source of income other than investment income, farming receipts from poor quality land, together with property rents, the ultimate demise of the Ashley Park mansion and estate was set in train.

The general education of the children was not neglected although it took place in the home, initially from visiting tutors. As Oxford University had recently rescinded its bar on Jews it was thought appropriate to prepare the boys for entry there and with this in mind a tutor was engaged in 1870 when Joseph was already aged fourteen, initially resident in university vacations, but was to become fully resident on his graduation. This was Arthur Read, the eldest son of Robert Read, a Reigate preparatory school proprietor and headmaster. He was at that time still an undergraduate at Queen's College, Oxford, attaining his B.A. degree in 1874 and was to have a major influence on two generations of Sassoons over a period of fifty two years, ended only by his death. He began as tutor to the four children and as the three boys all obtained places at Oxford University and Rachel later became the editor of two national Sunday newspapers, his initial task can be counted as successful. When Joseph became of age, Arthur Read additionally became his private secretary and confidant and close friend, a relationship only to be ended by Joseph's death over fifty years later. He fostered Joseph's interest in sport, particularly cricket and golf. Arthur Read in turn shared Joseph's pleasure in building up the collection of rare books for his library, he tutored the next generation of the family by which time he was recognised as a respected member of the family. He was for many years captain of Walton on Thames Cricket Club which played in Ashley Park, captain of Ashley Park Golf Club throughout its existence and along with Joseph was one of the founders of Burhill Golf Club and its first captain. He outlived all of the first generation of his pupils save for Rachel. On his death in 1922 he was buried in the cemetery of St Mary's Parish Church in Walton where his headstone can still be seen. His younger brother Walter W. Read was an outstanding amateur cricketer for Surrey over a twenty three year period from 1873 and captain of the County club in their then most successful period in the 1880s, as well as representing England both at home and on tour. He was also responsible for bringing many first class cricketers to play in Ashley Park, including, in the memories of both family and villagers, the legendary W.G. Grace. Many clubs throughout the country have claimed such a link but Ashley Park has more credence than most. Certainly W.L. Murdoch who captained the Australian cricket teams over a long period, later made his home in England for many years and was a great friend of W.G. played there on at least three occasions. Many county and test cricketers also played at the ground, thereby enjoying Joseph's hospitality.

Joseph, like his father was by nature shy and retiring but he diligently applied himself to his studies. He duly matriculated at the age of 18 and entered Christ Church, Oxford in the autumn of 1874 where he read for a pass degree which involved a selection of subjects from a prepared list. He studied, amongst other subjects, classical and modern history and languages which were to be of continuing interest to him throughout his life and in which he showed considerable proficiency. In later life he was to write poetry in classical Greek. He learned Persian, the language of his forebears and even a smattering of Welsh from one of the maids. He began to build on his father's collection of Hebrew books at a not inconsiderable cost. He obtained his BA degree in the summer of 1877 taking his MA after the normal lapse of time. At this time Elias, now running his own successful banking and trading company, stood down as a trustee of S.D.'s will to make way for Joseph's appointment in his stead.

Six months earlier he had celebrated his coming of age in some style as he took over as the squire of Ashley Park. Great festivities were held there with 350 of the poor of the village (more than a tenth of the total population) sitting down to lunch, consuming four sheep from the park, roasted whole and unlimited beer, with sports for the children following on. Dancing preceded dinner served at 11pm for Flora, Joseph, his brothers and sister together with prominent local people. Joseph's health was proposed by Mr Eves, a local businessman to which he responded. George Warner, a well known local corn dealer proposed Flora's health to which Alfred, then only a precocious sixteen-year-old was given the task of responding. Arthur Read proposed the prosperity of the local tradesmen to which George Luker, head of a local firm of builders and decorators, replied on their behalf. The guests on the short list of those admitted to the house commented favourably on the life sized paintings of members of the Sassoon family in the house, mainly the work of members of the Thornycroft family into which Alfred was later to marry, and on paintings by Rachel and Alfred. Thus Joseph entered into his substantial inheritance as the bells were pealed in St Mary's church. To mark the occasion Flora acquired a piece of land in Walton at the junction of the High Street and Ashley Terrace, later renamed Churchfield Road, on which she had built a Public Hall, for the use of the townspeople, which was opened in 1879 at a cost of £1,200. Within ten years modifications had to be made to provide adequate access for it to be licensed for use by members of the public.The building can still be seen to this day, the front alas hidden since the 1920s behind a shopfront which now proclaims the Superdrug shop. The building was successively used as a public library, meeting hall, concert hall, polling station and cinema but was sold when the Sassoons vacated Ashley Park.

Rachel's early life at Ashley Park followed a similar pattern to that of her brothers and as befitted a wealthy young lady who was expected to marry well, was more inclined towards music, painting and deportment rather than the more academic subjects. However Arthur Read's grounding in the English language and literature was to be to her benefit in her journalistic career. She was a rather serious young lady, small in stature like her mother, but she did not appear responsive to the advances of the young men whom she met at the social events which she attended. She much preferred to spend her evenings in London at the theatre and concerts or in visiting art galleries. Six months after Joseph's coming of age she appeared in public locally as part of the musical entertainment to mark the rehanging of the bells in the parish church, an early example of the interest which the family continued to take in the local church, despite their differing faith. At the time of reaching her majority no suitable Jewish suitor had appeared on the scene and there was family concern that she might not marry.

Alfred was of a very different temperament from his elder brother having an outgoing personality with much charm and had a strong appeal to the opposite sex from an early age. He enchanted his mother with his many talents and she spoiled him excessively. He appeared to be able to excel at everything to which he turned his hand ranging through painting, riding, cricket, ballroom dancing to playing the violin. In the latter Flora saw him as a potential concert performer and arranged for a series of lessons and purchased two Stradivarius violins for him, goodness knows why two! Even Reuben who himself enjoyed an extravagant lifestyle was moved to suggest that such expenditure was somewhat excessive. Like so many dilettantes he did not have the application to fulfil his early talent in any one field but he gained pleasure from playing the violin until the end of his foreshortened life. He matriculated at the age of eighteen and took up residence at Exeter College, Oxford in the autumn of 1879 but failed his responsions examination at the first attempt before passing at the second. He fell at the next hurdle, pass school moderations which were made up of fairly elementary questions on a range of arts and science subjects, which he also failed on two subsequent occasions before giving up hope of a degree and leaving the university at the end of 1881. However he seems to have enjoyed the university life to the full. Although he did not appear in the college cricket team he played the game to a good club standard, having been coached for the previous eight years by Arthur Read.

His indulgent mother was then persuaded to allow him to embark on the Grand Tour but even there did not complete the course having travelled no further than Paris where he set up in style and cut quite a dash, having come into his inheritance. The original £5,000 legacy from his father together with its accumulated income would, at his coming of age, have been worth over £300,000 at today's values. Sarah Bernhardt whose exotic company he enjoyed described him as 'the rich nabob, Alfred Sassoon'.

Frederick (known in the family as Bey), of whom little is known and who did not appear in several previously published family trees was a year younger than Alfred. He was always much under Alfred's influence, but he applied himself to his earlier studies with greater diligence and although younger, matriculated at the same time. Together they went up to Exeter College, Oxford in 1879. He was initially more successful than Alfred passing both his responsions and pass school moderations at the first attempt followed by the divinity part of 'Rudiments of faith and religion'. He did not take the honours examination at the end of 1881 and he left the university, as he joined, in the company of the more dominent Alfred, thereby failing to obtain his degree.

The previously referred to dispute arising from S.D.'s will must now be considered in more detail. When Joseph came of age, under the terms of his father's will he was to inherit the Ashley Park Estate, a property of over 375 acres. The trustees of the estate were Flora, Solomon who was S.D.'s brother and resident in India, Antonio Sefi who was a business colleague of S.D. and Joseph who had been appointed on attaining his majority in place of Elias who had asked to retire. Clearly Flora, supported by Joseph were the dominant trustees and they held the estate in trust for Joseph. The fact that they did not transfer the property into Joseph's name straight away suggests that they may well have already been advised that the terms of the will lacked precision.

A study of the will shows that S.D. clearly bequeathed the Ashley Park Estate and other assets to Joseph. However, when he attempted to define the Estate and its integral properties he stated that it was approximately 250 acres and he then defined the boundaries of the Estate with a lack of clarity. Simplified and substituting the present day road names, the boundaries could be construed as Oatlands Drive on the west side, Bridge Street and High Street on the north, Ashley Road on the east and Oatlands Chase on the south. There appears to have been no dispute regarding the mansion and its gardens.

At some point of time Alfred considered that he had been deprived of a part of his legacy, but did not pursue the matter until his break with his Flora following his marriage outside the faith in 1884. By this time Frederick had come of age and was persuaded by Alfred to join him in an application to the High Court to determine the interpretation of the will. In the High Court hearing in April of 1886 Mr. Justice Pearson determined that Alfred and Frederick were entitled to a share in the Ashley Park Estate and that arbitrators be appointed to determine the way in which the Estate should be divided.

The outcome of the arbitration given four months later was that Joseph should be entitled to Ashley Park House and all the land lying to the north west of Ashley Road, but that the land lying to the south east of that road should be divided between Alfred and Frederick by drawing a line from the mid point of Stompond Lane (roughly where the Links now adjoins it) to where Station Avenue and Ashley Park Road now meet. Alfred was to inherit the land adjoining Ashley Road and Frederick that adjoining Hersham Road. As there was a disparity in the acreage between the shares of the three brothers, Joseph and Frederick should each pay £50 to Alfred in compensation.

Whilst this judgement was no doubt legally sound it would have meant that there would be three separate estates, none of which would have been a viable entity. Fortunately there followed a series of events which were to prevent this from happening.

Alfred, who was by this time married and living in Kent, had no interest in owning unproductive land in Walton which yielded little or no income. To meet his expensive lifestyle he preferred the capital. Having made his point he quickly agreed to sell his newly established share of the Ashley Park Estate to his brother, Frederick, for £350 (approximately £17,500 at today's prices). Frederick now owned slightly less than half of the Ashley Park Estate but it is not clear as to what he proposed to do with it. As he was unmarried and was dividing his time between Ashley Park and Flora's house in Hove, his expenses would have been comparatively modest. He could therefore afford to take his time in deciding what he was going to do with his newly acquired land.

Before he had made this decision he was struck down by illness which today would probably have been diagnosed as cancer of the throat was to lead to his death within three years at the early age of twenty seven at Ashley Park in the presence of his brother Joseph. The condition was no doubt exacerbated by the family habit of continual heavy smoking. Shortly before his death, in defiance of medical advice to the contrary he was otter hunting in bad weather from which he caught a chill which was considered by his family to have accelerated his demise. Although he died at Ashley Park, in the period following the court case, his relationship with Joseph was not as close as previously and he spent some time in Hove with his mother, whilst retaining a pied a terre in the family home to which he returned from time to time. He had some years earlier appointed his mother Flora, his brother Joseph and his uncles Arthur and Solomon and his friend, Lionel Davis, who at the time of the 1881 census was shown as a visitor in Ashley Park House, as trustees of his will which left Flora to enjoy the interest from his estate for life and thereafter it would pass to Joseph. Frederick did not have much luck in his choice of trustees as both Arthur and Solomon decided not to act following the disputed court case. This left Flora, Joseph and Lionel Davis as trustees, but shortly thereafter Lionel Davis was then removed by the High Court as he had been declared bankrupt and had absconded, to be replaced by Antonio Sefi.

Flora's life interest in this land, however, yielded very little income so Joseph quickly entered into negotiations with his mother to purchase her life interest for £2,000, thereby restoring the Estate to one unit. It can be seen that the only significant beneficiaries of this dispute were Flora, and of course the lawyers, largely at Joseph's expense.

Chapter VI

Flora – The later years

Flora continued to live and to rule at Ashley Park for some years after Joseph attained his majority as he was not as yet married and the younger children had not attained their majorities. Frederick, the youngest of the family, came of age in 1883 and within two years the old order had changed. Alfred, shortly after his return from Paris, shocked his mother by marrying out of the Jewish faith. Knowing that this would arouse strong opposition the marriage was arranged and took place in secret without Flora's presence. He was the first of David's children or grandchildren to so marry and his mother was mortified, particularly in the light of her late husband's expressed wishes. Flora took refuge in the synagogue when she was told that the wedding had taken place and there she laid curses on any children of the marriage. She went into mourning for her "dead son" and forbade her other children to have anything to do with him. In particular Joseph was enjoined to see that Alfred never set foot in Ashley Park again. Alas Joseph obeyed these instructions to the letter, thus the two brothers drifted apart. Alfred was therefore unable to turn to him for help and advice when his marriage to Theresa was falling apart. Flora also instructed that Alfred's name was never to be mentioned in her presence. Only Rachel had the pluck to ignore her mother's edicts and kept in touch with Alfred and his family until his death and beyond which was to be a particular comfort to Alfred's widow and her children in later years.

Alfred's marriage outside the faith caused Flora and the Sassoons now resident in London to get together to ensure that Joseph would not follow in his footsteps. To this end, through the good offices of the synagogue, they urgently sought as a suitable candidate for his hand; one who came from a good Jewish family which could make a financial contribution as it was already apparent that Joseph was not going to earn a living in the family firm or indeed in commerce in any form. To meet these requirements they set their stall for Louise, daughter of Baron Horace Gunzburg, a banker, philanthropist, and bibliophile of St. Petersburg, who was the highly regarded head of a strictly orthodox wealthy Jewish family. Horace had acquired land there which he had donated for the building of a synagogue, as well as subscribing a quarter of the building costs and funding roughly one third of the running costs, thus he was held in high esteem in Jewish circles there. Surprisingly the synagogue which he financed has survived all the traumas of the last eighty years in Russia, but it was found to be in a rather run down condition when visited recently by Joseph's grandson, also named Joseph who is therefore one of Horace's great grandchildren.

The Gunzburgs were Bavarian financiers who had been ennobled some years previously by the Grand Duke of Hesse for whom they acted as bankers, taking their title from a town of the same name in Southern Germany, near Ulm. They subsequently settled in St. Petersburg following which the family fortunes were greatly enhanced during the Crimean War when they had the concession for selling vodka to the Russian troops there. This led them to become bankers to the Russian nobility and as a result Louise had been present in court dress at the coronation of Tsar Alexander III in the summer of 1882. Like many of their wealthy Russian contemporaries, the Gunzburgs divided their time between their mansions on Galernaya in St. Petersburg and their property in Paris and it was in the latter that Louise was born in 1862. Her mother died at the early age of 36 and Louise took over the running of her father's salon in her place. She spent a large part of he early life in Paris and her father had complained that she spent too much of her time away from Russia and might be seduced by the shallowness of Parisian life.

Flora in Ashley Park before leaving for Hove.

Despite the fact that marriage to Joseph would increase such absences the temptations of Walton on Thames were not considered as great as those of Paris! Accordingly no obstacles were put in the way of the wedding which was arranged to take place in St. Petersburg in the autumn of 1884. No doubt Horace had considered that a marriage between the Sassoon and Gunzburg families would be beneficial to their respective businesses but was probably not fully aware that Joseph and his brothers had cut themselves off from the Sassoon family business. This was to continue into the next generation and none of the offspring of Joseph and Louise were later to enter the business.

The compliant Joseph accordingly set out to meet his bride for the first time but there followed a period of concern in both Walton and St. Petersburg as he did not arrive there when expected. He was tracked down to Berlin, but whether the problem was laissez faire or cold feet is a matter of speculation but he was duly sent on his way and the marriage took place as planned.

Flora did not remain long at Ashley Park after Louise's arrival. As Flora's mother in law, and namesake, had a house in Brighton, as had her brothers-in-law Sir Albert and Reuben in addition to their London houses, it was natural that her thoughts turned there and she duly acquired a house in Hove at 37 Adelaide Crescent to which she retired. Alas her mother-in-law was only to live for a further two years. However the move proved a happy solution for both Flora and Louise. Flora was visited there from time to time by her youngest son, Frederick. His visits there were however cut short by his untimely death. Under the terms of his will she was to inherit a life interest in his estate which by then included that part of the Ashley Park Estate which lay to the south east of Ashley Road. This land yielded very little income but the splitting of the Estate was clearly a source of irritation to Joseph who immediately entered into discussions with his mother to buy out her life interest in order to restore the Estate to one piece as his father had wished when writing his will twenty years earlier. The price agreed, £2,000 (slightly less than £100,000 at today's prices) seemed a high one at the time, but in the event did not prove to be so as Flora outlived Joseph and he out of necessity had begun to sell these parts of the estate for development less than twenty years later, long before Flora's death.

Despite the growing disparity in their financial circumstances Flora in Hove was there able to keep in touch with other members of the family, in particular Reuben, as the effects of their earlier feud tailed off. She was variable in temper, sometimes talking entertainingly on books and the theatre, and at other times deriding the wealthy Rothschild friends of her brothers in law. She remained singularly unimpressed by Reuben's growing closeness to the Prince of Wales and in her own right became a well known figure in the neighbourhood, always elegantly dressed in black and carrying a rolled umbrella which she used to great effect. Whilst she was a formidable and strong minded character she was also given to unexpected acts of kindness, particularly to children. One such occasion on which she encountered poor children playing on forbidden ground led her to the acquisition of a park in Hove named St Anne's Wells Gardens which she threw open to the public and employed one of the children's out of work father as gardener there. She was easily offended and would cut even her close friends if the mood took her. Above all she was a character who made an impression on her adopted town where she continued to live until her death at the end of the Great War. Of her children only Rachel outlived her.

Chapter VII

Alfred and his family

Alfred as already seen was the first of his generation to marry, having enjoyed his time at Oxford and in Paris, following which he then returned to London. He had grown up approaching middle height and slim in build which gave him the appearance of being somewhat taller than he was. His social charms coupled with an inability to focus his talents had made it inevitable that he would not be tied down by the constraints of the family business, or indeed any business. Theresa Thornycroft who was born Georgiana but early on adopted the name Theresa became the subject of his attention. The Sassoon family had known the Thornycrofts since the 1860s when Mary (Theresa's mother), a talented artist and sculptor, had painted portraits of, and sculpted members of the Sassoon family at Ashley Park. Thereafter Theresa became friendly with Rachel and visited her at the Sassoon home on a number of occasions, a friendship which was to endure despite the difficult times that they were each to suffer, and was separated only by Rachel's death forty five years later. In 1882 Hamo Thornycroft (Theresa's brother) was commissioned by Flora Sassoon to make a half sized bust of her daughter, Rachel sitting on a music stool which he called 'A sonata of Beethoven' which was exhibited at the Royal Academy in the following year, but it was considered to be one of his least successful works. Hamo's wife, Agatha was later to describe it as 'graceful, but the face not pleasing'. Theresa had recently visited Paris with her brother Hamo to whom she was to remain close, and it was quite possible that she and Alfred renewed their earlier acquaintance there.

The Thornycroft family could trace their ancestry to landed farming stock in Cheshire. Theresa's grandfather John farmed Great Tidnock Farm at Gawsworth on the outskirts of Macclesfield. By the early part of the nineteenth century her father Thomas had migrated to London where he made a name for himself as a sculptor making an equestrian statue of Queen Victoria which was to be seen at the Great Exhibition of 1851. His other great interest was in mechanical engineering and the making of large model boats. His wife, Mary in addition to being an artist who produced a number of royal portraits was also a sculptress in her own right. They produced seven children, two sons and five daughters the youngest of whom was Theresa, who was brought up in Chiswick and became a student in the studio of Ford Maddox Brown where she met Nellie, later to marry Edmund Gosse who was to prove helpful to her son Seigfried in advancing his literary career during and after the Great War. The children all showed artistic talent, in some cases coupled with an interest in engineering. Her elder brother Hamo was a sculptor of note and perhaps his best known work is Boadeccia on her chariot which stands at the north end of Westminster Bridge opposite the Houses of Parliament. Theresa who was born in 1853 was no exception and painted to a high standard. Another brother, John inherited Thomas' interest in mechanical engineering, being apprenticed to a shipbuilder in the north east of England before going on to found, in his home town of Chiswick, a shipbuilding firm bearing the family name whose motor torpedo boats of his design were to play such an important role in The Great War and beyond.

Theresa was not the first of the Thornycroft family to have connections with Walton on Thames. Major General Roger Handasyde moved into Walton Lodge there with his wife Elizabeth, who was the daughter of Sir John Thornycroft Bart. of Thornycroft Hall in the County of Cheshire, not three miles from Theresa's grandfather's farm. The Major General did not live long in Walton but Elizabeth continued to live in Walton Lodge until her death in 1772 and lies buried in the Parish Church.

What attracted Alfred towards Theresa is very difficult to understand. He was very much a man about town, aged twenty two and thoroughly spoilt, whilst she was eight years his senior, somewhat unworldly and recovering from the break-up of an earlier romance. He was brought up in a strict Jewish household whereas she was an enthusiast of the High Church movement in the Church of England. They may well have shared a sense of humour, as Theresa was considered the humorist of the Thornycroft family; and they were certainly both of an artistic bent, he in music and she in painting and sculpture. Pressed by Theresa, Alfred enrolled as a sculpture student at the Royal Academy, but his powers of application were no greater there than at Oxford. Otherwise it was the attraction of opposites. Flora mounted a fearsome opposition to the idea of marriage but they were not to be dissuaded. They became engaged in the latter part of 1883, following which Theresa fled to Scotland to escape Flora's wrath while in great secrecy the wedding was arranged for January 1884. Fearing attempts to prevent the marriage, Theresa's brother Hamo was the only member of either family to be aware of the arrangements for which a special licence was obtained. The wedding duly took place at St. Mary Abbott's Church, Kensington, with Hamo thus being the only member of either family to support the bride and groom. Alfred's occupation was described on the marriage certificate with some licence as 'sculptor'. The other family members were advised after the event, the Thornycrofts accepting the union, but Flora was distraught. Neither Alfred or his children were to set foot again in Ashley Park as Joseph sadly followed his mother's directions never to allow him there again.

Alfred and Theresa after a period of house hunting bought a house in the Weald of Kent named Weirleigh at Brenchley some seven miles east of Tunbridge Wells, three months after the wedding. In the interim they lived with Theresa's parents in their substantial house in Kensington. At that time Brenchley was still a very rural part of the county, the London sprawl not yet having embraced that area, although the proximity of the railway would mean that its days of rural peace would be gradually eroded. It was a substantial house built on a hill some thirty years earlier with a central tower of sixty feet, and having nine bedrooms, three reception rooms, library and study. The grounds extended to some seven acres; a substantial property even if modest by Ashley Park standards. They initially employed three maids, a cook and gardeners whose numbers increased as the family grew.

For Alfred, Brenchley seemed from the outset to be an unlikely setting in which he could be content to live in the peace and quiet, whereas for Theresa, who set up her studio at the top of the house with uninterrupted views over the Weald to the South Downs beyond, it was idyllic. In the early years everything appeared to be set fair, their first child, Michael being born within a year to be quickly followed by Seigfried and later Hamo. The name Seigfried which had not been used previously by either family reflected Theresa's love of Wagner's operas. Alfred enjoyed considerable success in village cricket from which he derived much pleasure, playing regularly for nearby Matfield. He and Theresa installed tennis courts on which both they and their neighbours played and they stabled two hunters to enjoy riding together in the surrounding countryside, one of the few interests which they shared. Theresa was a fearless rider and she also enjoyed skating on frozen ponds in the colder winters then experienced. She even managed to encourage Alfred into the studio where he painted in water-colours the view from the window looking out over the Weald. Their son Seigfried writing some years later described his painting as uninhibited but amateurish. Theresa's art flourished during this period painting very ambitious religious subjects. She was particularly skilled in composition, her most successful picture was 'The Hours' in which twenty four figures float across the sky, passing from darkness into light. It was exhibited to acclaim at the Royal Academy in 1889.

Alfred, like his mother was unimpressed by the growing closeness of his uncles Albert, Reuben, and Arthur to the Royal family and to the Prince of Wales in particular and felt that they needed to be brought down to earth and reminded of their more humble background. He saw the opportunity to make his point during the State visit of the Shah of Persia in the summer of 1889. The Shah's previous visit having been something of a disaster, Sir Albert Sassoon was pressed by the Prince of Wales to act as unofficial host, having regard to his ability to converse with the Shah without recourse to an interpreter. Thus Albert's house in

Kensington Gore became the focal point of the visit attracting large crowds to watch the comings and goings of the nobility. There they were surprised one day to see a group of dirty, sinister looking Arabs arrive swearing at three camels which had clearly seen better days. As the camels sank to their knees sacks were unloaded at the front door, allegedly being presents from the Shah's loyal followers in England. Communication between the Arabs and the butler proved difficult and the party left apparently mouthing Arab oaths. The evil smelling sacks were found to contain only newspapers and dirty rags. The culprit was soon traced and proved to be Alfred who had employed resting actors and had hired the camels from a circus. The escapade, however, did no harm to Sir Albert who was made a baronet in the following year for his services, not unrelated to the success of this visit.

In the early years Theresa was an enthusiastic hostess as they entertained and their hospitality was well received locally. They in turn were welcomed in the large houses in the neighbourhood, where Alfred's musical prowess was much in demand. By 1890 it was apparent that Alfred was becoming bored with life at Weirleigh as he found that his older wife was content with the quiet life. He was now finding the excuse to spend ever more time in London. By 1891 matters had reached breaking point and Alfred acquired a small studio flat in Pembroke Gardens, South Kensington, an area much frequented by artists. To make matters worse for Theresa he was joined there for a short time by her best friend Constance Fletcher who had been a witness at their wedding. She was a novelist who wrote under the pen name George Fleming. Theresa in public put a brave face on the break up of the marriage and took solace in her growing boys, her painting and her horse riding. Her grief was real and she never gave up hope that Alfred would tire of his dalliance and return to the family home. His liaison with Constance was not to last but he then did not return to Weirleigh other than to visit the children, at first regularly, but the visits tailed off as his health worsened. These were extremely painful times for Theresa.

Seigfried writing some forty years later of one such visit in 1892 describes the three boys looking expectantly from the nursery window as Pappy, as they called him, alighted from the village fly at the gate. Excitement mounted as they heard his footsteps leading to the nursery and they crowded round him. As he kissed them his moustache tickled their faces and the smell of tobacco was unmistakable. He brought them presents of exotic foods and toys and together they romped on the floor whilst they laughed at his jokes. He had tea in the nursery while the boys stood around. Too soon for them he looked at his watch and took his leave, talking in conspiratorial undertones on his way out with their nurse Mrs Mitchell who ruled the nursery with a rod of iron. During his visit Theresa shut herself in the drawing room and they were careful not to meet.

Such visits became less frequent as Alfred's cough became steadily worse and by 1893 it was clear that he was seriously ill. Soon it became necessary for him to move from London to 30 Grand Parade, Eastbourne where the sea air was felt to be beneficial for his health. Early in 1894 Mrs Mitchell took the boys to see him in Eastbourne for several weeks. Alfred talked of getting well soon but he was so thin that his clothes hung from him and his face was transparent. Mrs Mitchell was unable to conceal from them that he was suffering from what she called 'galloping consumption', otherwise tuberculosis. It became all too clear that he was not going to recover.

In November of that year Mrs Mitchell again took the children to Eastbourne and on the journey told them that they would meet their paternal grandmother there. Flora had arranged to travel from Hove to Eastbourne to see their father, breaking her earlier vow never to see him again. There the children met her for the first time standing by the bed and speaking with a foreign accent. Their Uncle Joseph, his ban on meeting Alfred also having been lifted, was also there and spent most of his time gazing out of the window and said little. The boys were no longer permitted to kiss their father for fear of infection. Flora asked them many questions without appearing to take any notice of their answers which appears to have been

Alfred with his sons Michael, Seigfried and Hamo, c1908.

a Sassoon family trait. She then produced a scroll showing the family tree and pointed to their great grandfather at the top and themselves at the bottom. They had been surprised to see her as they had understood that she and their father had quarrelled and they no longer saw one another.

From the date of Alfred's will it is clear that Flora and Joseph's visit was connected with its making. In the will it was stated that Theresa had already been provided for in the marriage settlement and accordingly would receive only a legacy of £200 and personal possessions. The children would, on their majority, inherit the bulk of his estate which amounted to only £5,400 (£260,000 at today's values). This included properties in Walton on Thames, named Sassoon Cottages (thus named to this day) in Cottimore Crescent, together with two houses between Sidney Road and Hersham Road and seventeen acres of land in Terrace Road, adjoining Walton Grove. In later years, the trustees of Alfred's estate were to be involved in a court action over disputed access with the owner of Walton Grove, Paul Cababe a lawyer who needed little encouragement to resort to litigation. Mrs Mitchell was to receive £100 a year whilst she remained the children's nurse. In the light of her long-standing hostility towards Theresa which had recently become more open, it is not surprising that she would not remain their nurse for long after Alfred's death!

Alfred continued to struggle against the inevitable and finally died in April of the following year which was a happy release for him. Flora took over the funeral arrangements from which Theresa was totally excluded. Theresa had wished that he would be buried at Brenchley church but Flora decided against this and insisted that he should be buried in the Jewish cemetery in the East End of London alongside his father in one of the spaces which she had purchased there. Michael and Hamo attended the ceremony with Mrs Mitchell but Theresa was not invited. Seigfried was too distraught to join them. The boys found the whole event distressing, particularly the rabbis in their strange garb and the service conducted in a language which they did not understand. They recognised Joseph but none of the other mourners. They were consoled on their return by their mother whose grief must have been exacerbated by the deplorable treatment which she received at the hands of Alfred's family.

The sons of Alfred and Theresa are not really part of this story as they never lived in or visited Ashley Park during its ownership by the Sassoon family. Only Seigfried went there after the estate was sold, in the company of the novelist E.M. Forster in the course of a visit which he paid to his friend at his house in Weybridge in July 1923, only three months after the park had been sold. The house he then described as derelict and it would be pulled down shortly thereafter. He could only try to think of it as it must have been during his father's happy childhood, before an unsuccessful marriage and ill health together took their toll.

Given this tenuous connection with Ashley Park and that Seigfried was the best known of any of the descendants of the Ashley Park Sassoons, and that his writings give an unique insight on his parents' marriage and of Rachel's lifestyle, a brief outline of his life would seem appropriate.

Theresa continued to live at Weirleigh where the children enjoyed an idyllic childhood free from want as Theresa, helped financially by the Thornycroft family proved to be a prudent manager but nevertheless felt it necessary to sell off Alfred's jewellery from time to time to make ends meet. Aged thirteen Seigfried went to Marlborough College as a boarder but there he found sporting and rural pursuits to be more congenial than his academic studies. His final school report was very perceptive, stating "seems unlikely to adopt any special career". Theresa indulged these non-academic activities, the funds held in trust for the children even running to the buying of a hunter for him. Poetry then first began to interest him and it was decided that he should proceed to University but, influenced by the Thornycrofts, it was to Clare College, Cambridge that he went rather than to Oxford as his father had done. His lack of application there however matched that of his father and he left without a degree despite the strong pressure to the contrary of his uncle Hamo and his fellow trustee, the solicitor Herbert Lousada. After coming down he was to spend his time largely in enjoyment, living well on an income of £600 a year. The Great War was to change that for him as for many of his contemporaries.

He eagerly joined up, initially as a trooper in the Sussex Yeomanry but by May 1915 he was commissioned into the Royal Welch Fusiliers and with his regiment went to the front in France. There he experienced both the exhilaration and the despair of war which was to influence his poetry and his life. He proved to be a daring, not to say a foolhardy soldier, won the Military Cross in June 1916 on the Somme and was christened 'Mad Jack' by his regimental colleagues. Disillusioned, he was later to throw his medal ribbon into the River Mersey. Like many others he had lost confidence in the conduct of the War by the High Command whom he publicly criticised in 'A Soldier's Declaration'. There was a move for him to be court martialled, but to avoid the adverse publicity which would have followed, he was sent to a psychiatric hospital in Craiglockhart, Edinburgh run by an enlightened doctor – William Rivers where he was to meet his fellow poet Wilfred Owen. His poetry at this time was some of the most influential in highlighting the deficiencies in the conduct of the war which he published in a controversial volume of poems 'The Old Huntsman'. Recovering there from his supposed shell-shock, he soon wished to return to his regiment which he did during the last year of the war, firstly in Egypt and then once again on the Western Front and despite continued recklessness, resulting in further wounds, he survived the war unlike his brother Hamo whose death at Gallipoli in 1915 had severely affected him. Theresa never fully recovered from the shock of Hamo's death. She was later to gain some solace from spiritualism and continued to live on at Weirleigh until her death, aged 94 in 1947, following the end of the Second World War. Shortly after Seigfried's return to the front, his uncle Joseph of Ashley Park, who had not approved of his public criticism of the conduct of the war, was moved to pen a short poem to him dated 1st June 1917 which being the only known only communication between them is worthy of recording. Like much of Joseph's advice any criticism is muted and almost apologetic and it would be interesting to know what Seigfried thought of this poetic effort made by his uncle. At least he thought enough of it to have kept it amongst the many papers which survived him.

O piteous grave! – and black, untimely loss!
And laughing eyes, too soon acquaint with tears!
Still the loved greetings faintly steal across
The waste of years.

A hero's shade now takes the grand adieu;
His meed of praise in stately numbers ranged;
And 'there's a war in France, my kinsman true!'
And 'men are changed'

After the war there followed a period in Seigfried's life where his poetry did not have the same intensity and his popularity faded as he dabbled in journalism as a literary editor. A legacy from his Aunt Rachel who died in 1927 after many years in care enabled him thereafter to live in some comfort. His life changed for the better with the publication of his autobiographical portrayal of country life in the late nineteenth and early twentieth century in 'Memoirs of a Fox Hunting Man', followed by 'Memoirs of an Infantry Officer' and 'Sherston's Progress', all of which received literary acclaim. From that date his popularity recovered and thereafter his literary career flourished.

Seigfried, much to the surprise of his friends who were aware of his homosexual tendencies, married late in life, aged 47 to Hester Gatty. They produced a son George who is still alive and living in the Western Isles of Scotland. The marriage was not however a success and from London he retreated to Heytesbury House, a rambling property in Warminster, Wiltshire, not far from his old school, Marlborough. The house alas was destroyed by fire and is to be sold for development. There Seigfried continued to write and to play cricket with skill until he was well over sixty. He acquired the striking portrait of his favourite aunt,

Rachel Beer by H. Thadeus Jones on the death of Rachel's nurse, Miss Ross, who had bought the portrait at the auction following Rachel's death. It was then given pride of place in his library at Heytesbury House throughout the remainder of his time there. At aged seventy, having been born into the Jewish faith and dallied with the High Anglican church, he converted to the Roman Catholicism.

To make up for his earlier failure to take a degree at Cambridge University he received some sixty years later an honorary degree of D.Litt. from his father's university, Oxford. He died peacefully in September 1967, aged eighty one.

Chapter VIII

Rachel

Rachel was in many respects both the most remarkable, the most original, most able and latterly the most tragic of her generation of the Ashley Park Sassoons. Although short in stature she had grown into a sought after young lady with her contemporaries. She had a paler complexion than her brothers but with the dark eyes which were so common amongst the Sassoon family. Young men found her attractive in both appearance and personality. Always volatile, she had a quick wit and could be charming when the mood took her. The fact that she had a substantial private income from S.D.'s legacy no doubt increased her attraction to the opposite sex, but may well have led her to doubt her suitors' motives. She showed greater independence in breaking free from the constraints imposed by her mother as chatelaine of Ashley Park and managed to persuade Flora to allow her to undertake voluntary nursing at the Brompton Hospital for two years which was to broaden her horizons as she was then able to live in a London flat of her own. With her family background in India it is of interest that at this time she became an addict of Kipling's works. Whatever the reason Rachel had not met a young Jewish man who could meet both her own and Flora's standards by the time that her mother left Ashley Park for Hove when Rachel was aged twenty seven.

Freed from the constraints imposed by her mother's regime she soon met Frederick Beer who like her was small in stature. He had drooping eyes, a thick growth of beard and was of a sensitive and extremely retiring nature, but he was not of the Jewish faith. It was inevitable that Flora would once again mount vigorous opposition to marriage but Frederick's wealth and the fact that any children would be considered Jewish as this right passed through the female line, made her opposition less vehement. Also Rachel was a strong and determined opponent and the marriage was duly arranged for the summer of 1887. Villagers of Walton talked thereafter of having seen Rachel's portrait in the Ashley Park mansion turned to face the wall. Whilst this may be taken with some reserve as the local people had little or no access to the house and Flora had already left Ashley Park over two years earlier, the fact that there was family opposition to Rachel's marriage outside the faith was common talk in the village underlining the strength of feeling which existed within the family.

Frederick was the only son of Julius Beer who was born in Frankfurt, had settled in England and quickly made his fortune on the London Stock Exchange. Julius had purchased *The Observer* newspaper in 1870 more as a means of gaining respectability than either as a profit making venture or for the propagation of his political views. He had lived in considerable style in Portland Place, south of Regents Park and wintered on the French Riviera where he had died at the early age of forty three.

The pushiness that had helped him to make his fortune had not been well received in London society where he had entertained lavishly but had failed to gain the hoped for acceptance. He had married an opera singer and there were considerable doubts as to Frederick's legitimacy. Theresa described him to the young Seigfried as having 'a deficient family tree and no relations'. Possibly for this reason and for his lack of drive, Frederick did not measure up to his father's expectations. Julius considered him a weakling and treated him with cruelty and contempt, thus increasing his timidity and lack of self respect.

Despite threats of excommunication from Flora, Rachel was not to be dissuaded and they were married in 1887 at Holy Trinity Church in Sloane Street, close to her flat and workplace. This church was shortly thereafter to be pulled down as it had become too small for the growing congregation it then attracted. It was replaced by the present larger church on the same site. Rachel, surprisingly given her lack of vanity, saw fit to deduct five years from her age when signing the register. It was not in character for her to do so deliberately, perhaps it was early evidence of her increasing forgetfulness. It was not wholly unexpected, given the faith of the groom, that Flora with her past record would decide not to attend the ceremony. In her absence the register was signed by Helen Thornycroft, Theresa's elder sister who was given to dressing in manly attire, William and Catherine Gladstone, William Dicey the editor of *The Observer* newspaper, and two members of Frederick's family from Frankfurt. Catherine Gladstone stood in for the absent Flora at the reception. The newly wedded couple set up house in some style at 7 Chesterfield Gardens, off Curzon Street, Mayfair. Chesterfield Gardens was a very fashionable street of houses having been created following the sale of a part of the garden of Chesterfield House for development in 1870. Number 7 was one of five large four storied houses subsequently built on the eastern side of the new cul-de-sac. It boasted a most impressive Adam staircase allegedly built at a cost of £20,000. Frederick had inherited a substantial fortune on his father's death, together with *The Observer* newspaper and enjoyed an income of £20,000 a year (equivalent to almost £1 million at today's values). In addition Rachel in her own right was a wealthy young lady from her father's bequest.

Frederick was thin, almost shrunken and pathetically shy but with an easy-going disposition. One of his business associates described him as resembling a frightened mouse. He seemed to hide behind his beard, his handshake was limp and a cigar regularly drooped from his lips without it appearing to give him any sense of enjoyment. After their marriage it became apparent that Frederick's health was not good and he was initially understood to be suffering from migraine spending a considerable part of his time lounging in an overheated room surrounded by medicines prescribed by doctors and various quacks. The worst of these quacks were soon dismissed by Rachel who took over the supervision of his care, with her limited medical knowledge. They initially entertained lavishly both in Chesterfield Gardens and in their Richmond villa as befitted their station and there seemed at first to be some improvement in Frederick's health as he was brought out of his shell. However the curtains of his room remained drawn and Rachel's piano playing supplemented the prescribed sleeping draughts. This proved to be a temporary release as his moods continued to fluctuate alarmingly. His illness was described as a progressive general paralysis which at that time was often used to avoid the use of the word syphilis. From the description of his illness it seems certain that he was suffering from that disease which he may have inherited, hence Theresa's reference to his "bad heredity". Whether Rachel was aware of the true nature of his illness is not known, although her limited nursing experience would suggest that she should have at least sensed the possibility. In all probability she was unwilling to accept the obvious.

Frederick by now was neither well enough nor had the inclination to edit *The Observer* or to take any interest in it but Rachel, following on from her father's early interest in journalism, began to write articles for the paper. In due course her pioneering journalistic activities began to increase as she was bitten by the Fleet Street bug. She was appointed deputy editor but she thirsted to emerge from under the wing of the ailing proprietor. She next became editor, at least in name, but as she did not enjoy full independence, she felt inhibited. Overcoming prejudice against both her religion and her sex, she became a member of the Institute of Journalists and her period as editor was described by one of her colleagues as a 'delicious interlude in the life of the paper'.

During this period the Dreyfus case in Paris had attracted considerable comment in London, not least for its anti semitism. Dreyfus, a Jew and a serving officer in the French army was tried for passing secrets to Germany, found guilty of treason and imprisoned for life on Devil's Island. In 1898, some four years after his incarceration *The Observer*'s correspondent in Paris advised Rachel that a Major Esterhazy was prepared to confess to forging the bordereau setting out the military secrets passed to Germany by Dreyfus, a significant document which was a major factor leading to his conviction. Esterhazy was accordingly brought to London where he was interviewed by Rachel personally in the presence of a witness. He confessed that he had written the bordereau under the orders of his superior officer as the case required more than circumstantial evidence to ensure a conviction. Rachel, now being fully convinced of Esterhazy's truthfulness published two articles in *The Observer* which caused a furore world-wide with calls for a retrial, which was refused. Esterhazy subsequently retracted his confession and sued *The Observer* and its editor for libel. Rachel, represented by Sir George Lewis, known to her as the tenant of Ashley Cottage on her brother's estate, refused to back down and the case was eventually settled out of court at a cost to the paper of £500. In addition, the editor who shared with Dreyfus a Jewish background was thought by some to have acted from religious motives, but Rachel stood firm, thereby winning the respect of her fellow journalists. In 1899 Esterhazy finally admitted to his duplicity and Dreyfus was subsequently released, due in no small measure to Rachel's persistence.

When the Princess of Wales opened the press bazaar in the latter months of 1896 she complimented Rachel on the appearance of *The Observer*'s stall. Rachel had invited her nephews from Weirleigh to act as page boys at the event, suitably dressed up for the occasion. Seigfried, then aged ten, tripped as the boys were called to take their bow to the Princess and her escort, the formidable Duke of Cambridge. Rachel was not at all put out and suitably rewarded the boys, Seigfried getting a cricket bat from the Wisden shop which helped to whet his appetite for the game which lasted throughout his life.

As Frederick's health moved into terminal decline, Rachel's leaders in the newspaper became more irrational, including one commenting favourably on cannibalism, and her general behaviour became even more absent minded. The paper not surprisingly followed her health into decline and after earlier unsuccessful efforts to dispose of it, was sold on behalf of Frederick's estate to Alfred Harmsworth for £4,000. Unfortunately the manuscript of the history of the paper which existed at Frederick's death was later destroyed by Lord Northcliffe thereby depriving future generations of the full story of Rachel's editorial activities.

In parallel to her work for *The Observer*, where she did not initially enjoy full control, she sought an additional outlet for her journalism. In 1893 the opportunity arose to purchase its rival the *Sunday Times* and Rachel could not resist the temptation to acquire it at a cost of £11,000. The paper then cost a penny a copy for the eight pages which unimaginatively supported liberalism and was then in a poor financial state. Rachel's initial aim was not just to be a passive proprietor but to become its editor as well and to make the paper more objective thereby improving both its circulation and its financial position. Gladstone, although now a deaf and ailing old man, had recently become Prime Minister for the last time and she supported his domestic policies, but in foreign policy her leanings were much closer to those of Disraeli and his successor Lord Salisbury. In reality she was more of an editor in chief than a working editor, the editorial duties being mainly carried out by Arthur Beckett who combined the role with that of deputy editor of *Punch*. Rachel wrote many of the leaders under the heading 'The World's work' with the subheading 'What should they know of England who only England know?' Her editorials were described as having all the shrewdness and common-sense of her sex but would have been better for a little manly punch. As editor she was slapdash, her handwriting almost illegible, and her time keeping had become increasingly erratic. Their house became more and more submerged under books sent for review which were never read. From time to time the Sassoon boys at Weirleigh received entirely inappropriate books from this source as presents. As an example the young Seigfried, in his early teens, received as a Christmas present a copy of 'Bibliography of the World's Municipal Literature' which cannot have made the most exciting Christmas reading. An apprentice at the *Sunday Times*, Gus Wingrove whose duties at that time included the taking of proofs of the leader page to Rachel's house for her approval, described spending a considerable time waiting in the kitchen for the amended sheets. On the occasions when Rachel had dined particularly well the footman had to be sent to rouse her and the apprentice would then be despatched to his office in her open landau accompanied by both coachman and footman, in order to meet his deadline. She continued as editor until 1897 when, having lost interest she gave up. It would appear something of a miracle that the paper had appeared at all, let alone on a regular basis. T.P. O'Connor veteran editor and Member of Parliament had occasion to attempt to negotiate a business deal with Rachel and Frederick. He described them as a pathetically shy pair, he even more so than her, who he found 'not unattractive'. He said that it was almost impossible to carry on a business discussion with them. Writing in the *Sunday Times* after her death many years later he summed up her editorship with the words 'Altogether she made something of a hash of it'. The paper limped on until 1903 when it was bought by the brasher *Sunday Special*, founded six years previously, with which it merged whilst retaining the more established title.

Rachel continued to be ostracised by her mother for having married outside the faith and with Joseph taking his lead from Flora, she had become closer to Theresa as Alfred's health deteriorated. Her time keeping, or lack of it, was shocking and on one visit to Weirleigh she missed her train and hired a 'Special' at considerable cost, apparently a not infrequent happening. Arriving by the village fly, Seigfried described her as always looking out of place in the country with well cut London clothes put on in a hurry. He noticed that her hands were filthy and wondered when the rings on her fingers had last been removed in order to wash. In short she was an eccentric. After spending some time with the children she would enter into deep conversation with Theresa before rushing off as she said that she could not leave

Frederick for long, despite the presence of many servants to look after his needs in Chesterfield Gardens. It was clear that Frederick's health was continuing its downward track and he was becoming a total invalid but Rachel refused to accept the situation, even going to the length of buying cricket gear for him to practice at their Richmond villa!

Theresa and her children also made occasional visits to the Beer household in Chesterfield Gardens and writing in 'The Old Century and Seven more Years', in the 1930s Seigfried describes one such visit which gives an insight of the lifestyle of the household there. On arrival at Charing Cross they were met by the Beer family coachman, a large man attired in a long brown coat and cockaded hat. He escorted them to the luxurious brougham, bearing the family crest, which depicted a pelican feeding its young. This was harnessed to two sleek brown, well fed horses. On arrival at the massive portico in Chesterfield Gardens they were met by the footman, the image of the coachman, also dressed in brown, as were all the servants. Once through the doors with their guilded metal grille they were received by the butler and several more footmen who took their coats. The brown uniforms were everywhere with their gilt buttons bearing the family crest. Even the little black poodle had the family crest clipped out on its back. Escorted into the drawing room to await Rachel they were able to admire the many gilt Chinese cabinets and foggy antique mirrors and heavy damasks and brocades. The French clock tolled the quarter hours which served to emphasise her well known lateness. At last she appeared wearing a large feathery hat and smelling strongly of violets. Her conversation was vague and in such murmured tones that it was difficult to follow. Summoned by the gong, they passed along a glass walled corridor made to look like a bamboo bridge which was lit by electric light, then still a novelty. The dining room, also lit by electricity, nevertheless was gloomy as it had no windows, no doubt to protect the numerous valuable paintings which covered the walls from being damaged by sunlight. The lunch consisted of a number of courses, each more exotic than the last and Rachel, fuelled by the wine which she much enjoyed, became more relaxed and laughter broke through the gloom of the house. A visit to a matinee usually followed and to ensure that the children did not suffer from malnutrition further food was provided! This unaccustomed rich food often took its toll on the way home.

Rachel's nightmare existence nursing her dying husband finally ended in 1903 with his death. Only the intervention of Theresa enabled the funeral to take place as she had to prise Rachel from her room where the body lay, and make her take food. She also had to be persuaded to part with the body, which was eventually buried in the Beer family mausoleum in Highgate cemetery. Apart from Rachel, Theresa and her children there were few other mourners as he had been a reclusive invalid for many years. Rachel went to pieces completely following his death which she refused to accept. It became clear that her mind was going as she followed Frederick's downhill path, having presumably contracted syphilis from him. Within a year she was certified as insane and placed in the care of the Commissioners of Lunacy. As Frederick had died without making a will and Rachel was not capable of administering his estate, Letters of Administration were granted to her surviving brother, Joseph. Under the laws governing intestacy she inherited the whole of his estate, valued at over £400,000 (almost £20 million at today's prices. Only her wealth enabled her to avoid the indignities of an institution but it was necessary for her to have full time nursing care, and a more rural environment was advised by her doctors. Joseph was also appointed to act as administrator in her incapacity. It is interesting that in spite of the reclusive nature of their later years together when entertaining had become a thing of the past, Frederick's wine cellar when auctioned at Christies yielded in excess of £300 (in today's values over £14,000), which suggested that they had continued to take solace in drink. As can be seen, Rachel by any standards was a very wealthy lady which was further supplemented by the proceeds from the sale of Chesterfield Gardens. She moved to Chancellor House in Tunbridge Wells where she continued to live in some style.

Although not in London, this property was only marginally less imposing than the Mayfair mansion and dated from the latter part of the seventeenth century. Together with the nurses, Miss Ross and Miss Howard, the staff was on much the same scale as before with a number of servants moving with her from London. Even the family carriage was kept on with the coachman and footmen in their brown livery. Neither was she parted from her wonderful art collection which included a Canaletto, a Constable, a Hoppner, a Millais and no less than thirteen Corots, together with many others of note. The nursing care was beneficial and gradually brought about some improvement so that she regained her interest in the outside world. A splendid organ was installed in her drawing room which was to give her much pleasure. Theresa who lived nearby visited her from time to time which enabled her to follow the progress of her nephews. With the coming of the Great War she supported the Rusthall V A D and endowed a ward in the local hospital which was named in her honour. Rachel's rural peace in Tunbridge Wells was not immune to the sounds of war as in October 1915 a lone Zeppelin dropped three bombs on the town. Although no damage was done as the bombs fell on parkland there was panic among the local population who had considered themselves to be immune from such events.

The original open air concerts in Tunbridge Wells were financed by her and fetes were held in the grounds of the house as they had been in Ashley Park in her happier days. It was clear in the post war years that she was failing and it was no surprise when she died in 1927 having outlived all her generation of the Ashley Park Sassoons. The funeral took place in the borough cemetery, the service being conducted by the Church of England vicar of St. Pauls Church, Rusthall. The mourners included Theresa with her two surviving sons, Michael and Seigfried whilst Major Arthur Sassoon and his wife represented the Ashley Park branch of the Sassoon family. Her long time friend Ben Greest was also present.

At three auctions held at Christie's shortly after Rachel's death her jewellery realised over £4,700 and her paintings an impressive £23,600 (over £460,000 at today's values). Miss Ross purchased three paintings including the portrait of Rachel in a yellow dress holding an ostrich fan for 25 guineas. Seigfried had expressed interest in the painting but deferred to the wishes of Rachel's nurse and on her death it duly passed to him, to be hung in the place of honour in his house at Heytesbury. Like Frederick, Rachel had failed to make a will, in this case due to her incapacity. Letters of Administration were granted to two of her nephews, Arthur Sassoon, then living in Watford and Michael Sassoon who lived in Paddock Wood, not far away from Theresa at Weirleigh. Under the rules of intestacy her estate amounting to £277,000 (equivalent to nearly £4.5 million today) was divided between her six surviving nephews, one niece and one great niece (the late Totts' daughter Penelope) who each received approximately £30,000.

Chapter IX

Joseph, Louise and their young family

In July of 1880, at the young age of twenty four, Joseph obtained his first appointment in the adult world becoming a Justice of the Peace, following in his grandfather's footsteps and began to sit regularly on the County Bench in Kingston, learning his role. In later years he sat in court less often but continued to discharge the other functions with diligence, remaining active as a Justice of the Peace until shortly before his death thirty eight years later. This appointment crystallised his status high in the social hierarchy of the neighbourhood early in his adult life. A year later Joseph showed that he followed the political leanings of his father by being present at the formation meeting of the Walton, Weybridge, Cobham, Thames Ditton, Molesey and Esher Liberal Association held at the recently opened Public Hall in Walton where he seconded the motion to form the association. As Gladstone had recently become Prime Minister for the second time the support of S.D.'s son and heir was no doubt advised to him and welcomed. Joseph was not however overtly political, as he did not seek to influence others and preferring to keep his views to himself, but he remained of a liberal frame of mind throughout his life.

Bust of Louise as a young girl.

Following his marriage Joseph duly returned with his pretty young bride to Ashley Park at the end of their leisurely honeymoon on the Continent. She brought with her a bust in her likeness sculpted by R. Cauer, which is now in the possession of her grandson, Joseph. She quickly proved to be a strong willed and dominant young lady as had been the Sassoon wives of the previous generations. With the early death of her mother she had become her

Joseph and Louise.

father's favourite and as such was accustomed to getting her own way so it was inevitable that there would be a clash between her and Flora, who had become even more strong willed in the seventeen years of her widowhood as the undisputed ruler of Ashley Park. As befitted her Middle East background Flora expected to continue as head of the family, whereas Louise as Joseph's wife with her European upbringing had similar expectations. It soon became clear that these two strong minded ladies would not live easily together on the same estate let alone under the same roof. Joseph who had been dominated by his mother since S.D.'s premature death and who throughout his life shrank from confrontation was not initially disposed to forcing Flora out and as a compromise Flora retreated to Ashley Cottage on the estate.

In the summer of the following year the Gunzburg family, led by the Baron and accompanied by several of Louise's brothers descended on Ashley Park, the visit being arranged to welcome Joseph and Louise's first child, and to see how Louise was settling into her new surroundings. They explored the Park extensively, where they were photographed in somewhat artificial poses. The visit also enabled Horace to add his considerable weight to the pressure on Flora to move from Walton, to enable Louise to assume her rightful place in the Ashley Park heirarchy. Flora, who held the Baron in high esteem, found his reasoning persuasive and recognised that she had met her match. The Gunzburgs were to be regular visitors to Ashley Park over the next few years as life in Russia became more turbulent.

One of Joseph's sons described his father many years later as shy and retiring until he knew people very well and this description is echoed in the memories of the villagers who found that after the initial formal ceremonies had taken place at social or sporting events in the grounds of Ashley Park, he quickly disappeared from sight. He appeared terrified of appearing to influence others in any way. If he was persuaded to give advice it was done in few words and he would then turn away as if ashamed. He had a rather quirky sense of humour which he was convinced provided a necessary shield from the harsh outside world from which he was to increasingly shrink. He sincerely loved his home, his wife and his growing family but as the years advanced he retreated more and more into his family circle. Throughout his adult life he followed the family tradition as a generous subscriber to good causes, both national and local, not confined to Jewish charities. This was shown by his regular contributions to various appeals for funds from St. Mary's Parish Church in Walton and other local charities.

Louise relished her future as chatelaine of a substantial estate as she had a much more outgoing personality than Joseph and looked forward to taking part in the

Gunzburg and Sassoon families.

wider life of the local community. However the old Tudor house was showing signs of its age and she found the chilling draughts in the house to be even more numbing than the extreme cold of St. Petersburg where her father's substantial townhouse, a more modern building which was built to withstand the extremes of the Russian climate. Mains water and a gas supply was by now available and had improved the quality of life at Ashley Park and whilst Joseph had more than sufficient funds at this time to maintain the house, in itself a costly exercise, further major improvements he did not consider to be financially prudent, so Louise had to put up with some discomfort. She took to her role well, was highly thought of in the village as she gave garden parties and organised fetes and took her place in the social fabric of the neighbourhood. She began to entertain with enthusiasm within their close knit group of friends but Joseph's inherent shyness would restrain any significant broadening of this aspect of their activities. An example of the effect of Louise's personality on those she met is illustrated by the admission of Mr Justerini of the wine merchants, Justerini & Brooks whose Company supplied Ashley Park with wines and spirits during the Sassoon era "That he had been in love with Louise for most of his life but had never told her so."

Unhappily Joseph continued to feel bound by his promise to Flora not to admit Alfred or his family to Ashley Park as it was quite probable that Louise and Theresa would have found many interests in common to the advantage of both families. The rift was later widened when Alfred, supported by Frederick, successfully challenged Joseph's inheritance of the whole of the land comprising the Ashley Park Estate. As has been seen the result was that Joseph had to cede approximately half of his land, albeit of poor quality, to his brothers,which he was later able to repurchase. When Rachel, having also married outside the faith, suffered a similar fate to Alfred, she too sadly was never to grow close to her sister in law Louise. Rachel, with her own time increasingly taken up with her ailing husband and her editorial work, scarcely had time to be concerned about her brother's action, but when her mind gave way Joseph was her logical legal guardian and he did not let her down, although there is little evidence to suggest that he was more than an occasional visitor to her in Tunbridge Wells, other than out of necessity to fulfil his legal role. Of course by this time Frederick Beer was already dead which may have made Flora's opposition slightly less intransigent.

A year after Joseph and Louise were married, their first son was born and was named Sassoon Joseph following the pattern started by his grandfather in using the family name followed by the father's forename. Unreasonably burdened with this mouthful it was natural that within the family he should be known by a more informal name. In the family the name Totts duly emerged, although the reasons for it are not known by the present generation of Sassoons, all of whom knew him only by that name. To avoid confusion he will be referred to as Totts in this book. This event was followed by the birth of Hannah two years later, known in the family as Missy and thereafter with only yearly gaps, David, Arthur and Mozelle. The family was completed by the birth of twins, Edward and Frederick, known as Teddie and Freddie, in 1892. This rapid growth in family necessitated significant increases in the indoor staff and the census of 1891 shows eleven staff living in the main house comprising a private secretary (Arthur Read), two footmen, a nurse and two nursery maids, three house maids and two kitchen maids. In addition there were thirteen staff living in other premises on the estate, together with their families. No doubt the arrival of the twins in the following year would have caused a further increase, not least in the nursery staff.

It is possible to gain an insight into life in Ashley Park as the children were growing up from memories noted down by Joseph's second son David late in his

The twins Freddie and Teddie. The sash indicates that Teddie has been fed.

The seven Sassoon children, Mozelle, Missy, Freddie, Arthur, Totts, Teddie and David.

long life in which he aptly referred to those years as 'A vanished civilisation'. From this it is clear that the children had little or no contact with the village children and were consequently thrown together despite the disparity in their ages. Led by their tutor Arthur Read, who clearly saw it as a form of education, they formed 'The Republic of Ashlesia' which had a population of twelve, made up of seven children, Joseph, Louise, Arthur Read plus the two nursemaids. Arthur Read, despite being a confirmed royalist, was elected President and known as Five. (President Faure had recently become the President of France!) A pun typical of Joseph's sense of humour. Cabinet meetings were held weekly, minutes were taken and records were kept of the proceedings. They set up a Court of law with the President as judge, before whom any of the children could be brought for any misdemeanour. They had their own currency which was a rare fir cone found beneath only one tree in the Park (a weeping Norway Spruce). They traded goods for currency, formed a Chamber of Commerce and had a system of fines. Thus the children learned about conduct of meetings, court procedure and punishments, finance and current affairs. In addition Arthur Read taught them some elementary Latin and they learned a smattering of French from their mother. Their lack of a more formal early education would later make it difficult for them to shine in the more competitive environment of Eton College at which establishment the boys would in due course complete their school studies.

As befitted families of their class, the children were mainly looked after by nursemaids, in their case two sisters, Mabel and Maud whose high spiritedness matched that of their charges. However, over them was a head nurse, considered by the children to be a fiend incarnate, to whom they gave the name Grussy. She had a foul temper so the children took to the woods and shrubbery to escape her wrath. A bush telegraph system was instituted with the help of the nursemaids to warn of her impending approach. She was described as an artist with the slipper using the flexible sole late at night so that the marks were gone by the following morning. The two nursemaids smuggled in much-liked foodstuffs which were on the forbidden list. As soon as their parents were seen to depart in the carriage along the short avenue towards the station, Maud would run off down the long avenue to the village to buy hot buns from the bakers oven whilst Mabel acquired forbidden fruits from the estate. The coachman would be enlisted to buy further goodies after dropping off the parents at the station. Altogether the two maids were looked back upon with much affection by all the children in later life. As a matter of course all the children learned to ride, a white pony named Snowball being their first mount.

Arthur Read, Louise and Joseph on steps to the flower garden.

Arthur Read, Joseph (3rd), Louise (4th), and friends on the frozen lake, 1889.

Each summer Louise's father Horace took over a large house in Boulogne to which Joseph, Louise, the Ashley Park children and Mabel the nursemaid crossed by the packet steamer where they were joined by Louise's brothers and sisters and their families. Whilst there, on one occasion Mabel in resisting the unwelcome advances of a gendarme, emptied a bucket of water over him from an upstairs window. The gendarme wished to arrest her and returned with a colleague and a long argument followed with the Baron. It was agreed that the officers would be placated with a small financial consideration provided that Mabel apologised and agreed to behave in the future. This the high spirited girl refused to do. Further argument ensued before the officers were finally bought off at a considerable cost. Mabel, expecting the worst, was surprised to find only a light rebuke as her attitude was applauded by all concerned, particularly the children.

Ashley Park, as indeed Walton generally, was comparatively flat. The exception was a man made mound towards the west of the estate. This was the ice house which to this day survives in a private garden. It was created when a dowser found water and a well was dug 300 feet deep. Nearby a pond was created and puddled with clay to make it water tight. The ice which formed in the winter was cut and together with gathered snow was stored in the ice house for later use during the warmer weather. This enabled the provisions in the house to be maintained at a cool temperature throughout the year. As the ice house was situated at the highest part of the estate a water tank was installed on the top of it raising its height above the roof of the main house some 400 yards away. The tank was filled by pumping water from the well and the house supplied by gravity feed prior to the arrival of the public water supply in the last decade of the nineteenth century. The ice house chamber is in the shape of an inverted cone with a segmented dome entered by a tunnel nine feet in length which was out of bounds to the children for whom it held endless fascination, so inevitably they found ways of playing there undetected. As in so many houses of that time the servants would generally turn a blind eye to the forbidden activities of the children within reason.

Most evenings Louise, but on occasion Arthur Read, would read aloud from a book to the children with an occasional interruption or gesture from Joseph. At the end of the reading their father would tell them an anecdote, always instructive and told with humour, which all the children eagerly awaited, but which the younger ones often did not fully understand. The children enjoyed these readings and

Entrance to the ice house, Ashley Park. *(Elmbridge Museum)*

Joseph by the summer house, Ashley Park.

looked forward to these regular get togethers with their parents which were somewhat unusual in such families of that period.

Joseph kept loosely in touch with his relations in India, mainly through his uncles in London, particularly Albert and Reuben. His son David recalled an occasion during his early childhood when he was with his mother as the butler announced that there was an Indian man named Francis to see her. "Then show him in," she said. Then a tall, dark skinned man entered. "My name is Francis and I have come to cook for you," he announced at which she turned to the butler and said "Please show Francis to the kitchen." It turned out that he was a Goan cook, named Lazarus Francis, who had worked for the family in India and had been sent by them as a result of comments by Louise highlighting the difficulties in obtaining good cooks in England. One can only speculate on the reception which he received on his unexpected arrival in the kitchen. However he cannot have been too badly received as he was immediately employed and continued to work at Ashley Park until his retirement at the age of 65 and thereafter lived out the remaining years of his life in Walton until his death in 1903. In the notice of his death in the local newspaper he was described as a 'black man, conspicuous in the town' which at that time must have been a masterpiece of understatement. David's elder son, Joseph, the grandson of Joseph and Louise, has to this day copies of some of his recipes which he described as "Obviously intended for a large household and generous with the ingredients."

In 1894 there was the first public advertisement of the sale of plots for building of houses on the estate. The plots were situated in Station Avenue between Ashley Park Hotel and Ashley Road. They were described as "grand sites for first class houses on the Ashley Estate".

The new Church Hall adjoining St. Mary's Parish Church was opened at the end of 1903 by the Duchess of Albany, the wife of Prince Leopold, Queen Victoria's youngest son who lived at Claremont in Esher. Despite their Jewish faith Joseph, Louise and their family were in attendance, Joseph who had subscribed to the fund raising appeal being a member of the official reception committee.

Lazarus Francis, the Goan cook at Ashley Park.

Chapter X

Education of the children

This period of early childhood, so much enjoyed in the family was gradually coming to an end. Joseph had been persuaded that it was no longer sensible for the children's secondary education to take place wholly within the home as had been the case in his own childhood and that of his brothers. As something of an experiment, but not without trepidation, it was decided to send Totts to a public school and Eton was chosen for his further education despite there being no prior family connection with the school. Joseph would probably have been aware that his cousin Edward was contemplating sending his son Philip there. Totts duly started his education at the college in the autumn of 1898 and entered the house of Rev 'Hoppy' Damon, an Eton character approaching the end of his time there. He worked hard and made good academic progress through the school until his final year when his results fell back. He was always considered by his parents to be the most academic of the sons by a considerable margin, which was proved in due course to be a correct assessment. In sport he was a 'wet bob' making considerable progress in rowing in his last two years but narrowly failing to get into the college eight. He was to continue his interest in rowing, competing with some success in the Walton Regatta for a number of years.

In 1900 together with his fellow pupils at the college he lined the route as the whole nation celebrated Queen Victoria's eightieth birthday. Within two years he would again be called on to line the final lap of her funeral procession between the park gates and Windsor Castle. During his penultimate year at Eton his school house was gutted by a serious fire which resulted in two fatalities but neither Totts nor his brother Arthur, who was in his first term in the same house at the time, suffered any ill effects. Totts left the college for university in the summer of 1904 bound for his father's college at Oxford. There he sat papers in a modern language (probably German), elements of law and military subjects, graduating with a BA degree in 1907. By that time the military build up had already begun as concern mounted over Germany's growing naval and military might. Totts participated fully in the activities of the Officer Cadet Corps at the university, but showed no interest in business and as a result opted for a military career. As he was already an accomplished horseman with the right family background it was to be expected that he would choose to join a cavalry regiment. On coming down he therefore joined the 6th Inniskilling Dragoon Guards on a university commission. The regiment had amongst its battle honours Balaclava, having taken a major role in the Charge of the Heavy Brigade there. Unlike the better known but failed, Charge by the Light Brigade, the Heavy Brigade, although greatly outnumbered, successfully drove the entrenched Russian force from the heights. He had enjoyed his time at Eton and did his brothers a great service by convincing his father of the advantages of educating his brothers away from the narrow confines of Ashley Park. As a result they were all to follow him to Eton in due course.

Louise attended the Eton v. Harrow cricket match at Lords in the summer of 1901 in the company of Totts, in his second summer at Eton together with David then about to join the school and the younger boys who were to follow in due course.

She was described by another member of the Sassoon family present at the match as being "tightly laced into a concoction of Worth tulle." Not the most comfortable attire but very elegant.

The second son, David, duly joined his elder brother at Eton in the autumn of 1901 in the same house, which was now run by Mr R.S. Kindersley, who was described as a hard driving mathematics tutor. David did not enjoy good health during his time there and the trouble was eventually diagnosed as arising from a weak heart. The somewhat spartan life at the college was not considered by his doctor to be appropriate for his condition, and on that advice he left the college at Easter 1903 to take life more easily at home. On returning to his family at Ashley Park, he regained his strength and as his health improved he began to seek an interest to occupy his time which he found in landscape painting for which there was ample opportunity in the Park. By the autumn of 1906 his health had showed a marked improvement but it was felt that he should avoid the pressures of a boarding school education. To further his artistic leanings he was enrolled at the Slade School of Art, a college of London University. The school had been opened in 1871 on the strength of a legacy from Felix Slade but it was not until Frederick Brown was appointed professor in 1893, shortly to be joined as his assistant by Henry Tonks, that the school gained universal acclaim. David studied Fine Art and was much influenced by Tonks, both in his painting and by being introduced into the wider artistic community. Tonks surprisingly, as well as being the assistant professor, was a Fellow of the Royal College of Surgeons and was to succeed Brown as Professor in 1919. Through Tonks, David met Augustus John, a recent graduate from the school who was still much in evidence there. Thus commenced a life-long friendship between the two. Among other students who were David's contemporaries there were Stanley Spencer and Graham Sutherland. In 1910 whilst still a pupil at the Slade, David moved out of Ashley Park into the artistic world in London where he rented a small studio in Onslow Gardens, off the Fulham Road and lived well on an allowance from his father of £250 a year. He completed his studies in London in the summer of the following year.

Arthur followed his elder brothers into the same house at Eton in the autumn of 1902. His cousin Philip, grandson of Sir Albert, was his contemporary, although not in the same house, but they were dissimilar in almost all respects and saw little of one another. Arthur's academic progress was slow and it was unsurprising that as soon as he could he opted for the newly established Army Class at the beginning of the summer half of 1905. He played cricket for his house without distinction and left the college in the summer of 1907. He followed Totts in deciding to pursue a military career and entered the Royal Military College, Sandhurst in January 1908. In order to increase the throughput of officers to meet the needs of the growing army, the course had recently been reduced to a one year's duration so that he completed it in the December of the same year, being placed 98th in the order of merit and was gazetted to another cavalry regiment, 13th Hussars in January 1909.

Joseph was now fully convinced of the advantages of education outside the home so that the twins Freddie and Teddie, built on Arthur Read's early tuition, by entering a preparatory school in Sandy Lane, Cobham some five miles from Ashley Park. The school was Sandroyd Preparatory School which had established a reputation for preparing pupils for Eton under the headmastership of Mr. Wilson. The twins joined at the beginning of 1905 and were guided towards the entrance examination for Eton. The school flourished in Cobham until its evacuation to the West Country during World War II, where it thrives to this day. Freddie's sons Humphrey, Richard and John were to follow him there in the 1930s en route for Stowe. After the Second World War, Reeds School acquired the former site in Cobham, moving there from London and to this day it occupies the original Sandroyd buildings, now much enlarged. From there the twins followed the now well established route to Eton in the summer of 1906. An idea of the spartan conditions at Ashley Park at that time can be gained from one incident when Freddie was chastised for being late for dinner. In his defence he explained that as ice had formed in the wash hand basin he had been unable to do up his front collar stud due to frozen fingers. No doubt the conditions at Eton did not come as great a shock to the Sassoon boys as it did for other more pampered pupils. As befitted their preparatory school education, their early academic progress was somewhat better than that of their older brothers, but they too opted for the Army Class at the earliest opportunity. Both played cricket and football for the house, but they progressed no further on the sports field, although Freddie who was short and wiry was a notable schoolboy boxer at the lightest of weights. Together they left the college at Christmas 1910. Freddie showed promise at mathematics, defying family pressure to go on to Sandhurst and the cavalry, choosing instead to join the Royal Military Academy, Woolwich. The course was then of eighteen months duration, leading to a future role in the Royal Artillery, into which he was duly

Teddie and Freddie dressed for Eton College.

commissioned in July 1912. Edward followed the route of his brothers to Sandhurst and completed his course at the end of 1911, being gazetted to yet another cavalry regiment, the 5th Princess Charlotte of Wales' Dragoon Guards in the February of the following year.

The girls, Missie and Mozelle did not enjoy the same freedom as their brothers, being educated mainly at home. Missie, like David did not enjoy good health, suffering from time to time from a kidney complaint. She was described by one of the villagers as white faced, and dark haired and rather fat. Early photographs do not suggest that she was a fat child, but no doubt as her kidney condition worsened she put on weight. Her health caused increasing concern in the family and on medical advice she was sent to Hindhead for rest and recuperation but she died there of kidney failure at the Beacon Hotel in 1910 at the early age of twenty three in the presence of her brother David. She was buried with her Sassoon grandparents in the Jewish cemetery in Mile End in the presence of the family.

Mozelle latterly went to a local school run by a Doctor Dawes. Her father's only advice on her first day at school which showed his somewhat off-beat sense of humour as well as his knowledge of Shakespeare was "Don't wear your heart on your sleeve for 'daws to peck at"! The girls in particular suffered greatly from the coldness of the house in winter. After dinner as Joseph and Louise would sit near the fire in their Chippendale arm chairs embroidered on the back of which were the banners used at the funeral of Viscount Shannon (a previous occupant of Ashley Park who died in 1740). Arthur Read usually sat with them and the children were arrayed at the back of the gas lit room where the girls, dressed for dinner in their low cut gowns, felt the cold the most severely.

Chapter XI

Sport at Ashley Park

As well as being a scholar Arthur Read was a keen all round sportsman and was instrumental in encouraging Joseph and his sons and brothers to take an active interest in various sporting activities which were to provide them with much enjoyment throughout their lives. This interest in sport also provided the opportunity for Joseph to meet with the local people in whose company he was able to relax in the less formal sporting surroundings. Joseph was for many years a steward of Walton Regatta which was first held in 1862, in which his eldest son, Totts competed on a number of occasions around the turn of the century. At the regatta of 1913 Edward was able to arrange for his regimental band (5th Dragoon Guards) to play during the afternoon's festivities.

Joseph's earliest recorded participation in the sporting field was on the cricket pitch. As was the case in many landed estates across the country during the latter half of the nineteenth century, the landowner caused a cricket pitch to be laid out

Cricket party Lyrics v. Edgar Bruce's XI. Mr and Mrs Bruce (centre), Joseph (behind wearing boater), Arthur Read (to his left), W.W. Read (sitting) and W.L. Murdoch (to the right of Mrs Bruce)

on his estate to which the hospitality there dispensed enticed a wide range of prominent cricketers to participate in the delights of 'country house cricket'. Joseph was keen to join this trend and instructed his gardener to prepare a cricket pitch and to lay out a cricket ground in Ashley Park. Thus the game came to be enjoyed by the family, their guests, the servants and the villagers who mixed freely both on the field of play and in the small pavilion thereafter. G.M. Trevelyan, the historian, expressed the opinion that the free mingling of the classes on the cricket field and other similar occasions was the major reason why the French Revolution had not spread to England. Whilst it is not possible to be categorical as to the date on which cricket was first played on the ground. There is a newspaper report of a match between the gardeners of Walton and the gardeners of Kingston which took place on 25th August 1886 in Ashley Park by permission of Joseph Sassoon. There is clear evidence of the Walton on Thames Cricket club playing there regularly at the start of the next decade, with Joseph as its President, playing occasionally and Arthur Read as Captain more regularly. The pitch had been laid out on meadow land and judging by the early scores which survive was like many of its contemporaries of a 'sporting' nature, although being on sandy soil it was never dangerous. By the year 1900 under the stewardship of groundsman Burchett the pitch had improved to the extent that in a match between the Lyric Club and Edgar Bruce's XI, involving several test and county cricketers produced no less than 536 runs in the day. Edgar Bruce was a well known theatrical actor/manager and local resident whose house, The Beeches in Oatlands Drive, backed on to the estate. Cricket continued to be played without a break on the ground during the Sassoon ownership until sport was suspended during the Great War when such land was given over to agriculture. The ground was not restored prior to the sale of the estate by the family in 1923. Thereafter that part of the estate was acquired by the local council who restored the ground which exists on the same site to this day and is still the home of Walton on Thames Cricket Club. Despite over a hundred years of care and attention, the pitch retains many of the traits from its origins on the sandy soil.

Through Joseph's hospitality, coupled with Arthur Read's connections, the ground from its inception hosted many test and county cricketers of distinction, including W L Murdoch, captain of the Australian cricket team on their tour of England in 1890. After the tour had ended he stayed on and made his home in England, where he played county cricket for Sussex. Walter W. Read (Arthur Read's brother) was to be captain of the England test team and was for many years also captain of the Surrey County Cricket Club in one of their most successful periods. From their father's mouths the present generation of the Sassoon family have understood that the legendary W.G. Grace also played both cricket and golf on the estate and in the memories of old Walton families such stories have also been passed down. No verification of this has been possible and many grounds claimed this distinction. W.G. was certainly a friend and playing colleague of both Walter Read and Billy Murdoch who were responsible for bringing many prominent cricketers to Ashley Park. In addition W.G. was a keen golfer who rarely turned down the opportunity to play both sports on the same day even in comparative old age and Ashley Park provided such an opportunity. It is therefore highly likely that he did indeed play there. Arthur Read was the captain of Walton on Thames Cricket Club from its formation until 1911 when he was over sixty and still played, but not with the regularity of old. Joseph found that he could relax amongst the cricketers of the village and as well as being President of the Club, Joseph played a few games each season, mostly without distinction, until the mid 1890s by which time golf had taken over. He also enjoyed hosting the visits to his house and grounds by well known sportsmen who appeared to be able to bring him out of his shell.

Ashley Cottage (formerly the estate Ranger's house), a substantial property owned by Joseph in Oatlands Drive, following Flora's departure was let to the prominent solicitor George Lewis until 1905. His main residence was in Portland Place, London and he used the cottage as a summer residence. As a patron of the arts he entertained there such artists as Burne-Jones, Sargent and Alma Tadema. He also allowed Lily Langtry to stay there at a low point in her fortunes. He acted as solicitor to the Prince of Wales and many others of note and was knighted in 1892 and made a baronet ten years later, following the Prince's accession to the throne as Edward VII. Whilst staying at the cottage Edward Burne-Jones painted a portrait of Sir George's youngest daughter, Katie in a recumbent pose, one of his better known works.

The next tenant of note to occupy the cottage was the cricketer, F.R. Spofforth, one of the greatest Australian fast bowlers, who toured England with the Australian cricket teams of 1878 and 1882, earning the description 'the demon'. He subsequently married an English girl and settled in Walton from 1908 until shortly after Joseph's death. It is therefore more than probable that local cricketers playing in Ashley Park were from time to time observed at play by a man recognised as one of the greatest fast bowlers of his time.

Encouraged by his enjoyment of the cricket scene, Joseph took a great interest in the growing popularity of golf, a sport which he felt he would be able continue playing as he got older. As with cricket, he considered that there was sufficient ground on the estate to lay out a golf course, even if only of nine holes. The ground chosen for the course was that recently recovered from his brothers lying between Ashley Road, Station Avenue, Hersham Road and Stompond Lane. To implement

Louise golfing in Ashley Park.

his plan he turned to his gardener, James Albert Steer who had come to Ashley Park from one of the Royal Gardens. He was then aged twenty nine and proved to be an inspired choice. He laid out the course during 1890, using only manpower and horsepower. The course which was of nine holes which varied in length between 120 yards and 410 yards. The hazards were said to include trees, ditches and bunkers. The Ashley Park Golf Club was formed and affiliated to the Royal and Ancient Golf Club of St. Andrews. The course opened for play in the autumn of that year when James Steer became its first professional as well as greenkeeper. As was then common practice, no play was allowed on a Sunday before1 pm. even though Sunday was not a day of significance for its Jewish owner. The children's ponies were often pressed into service to pull the mowers to improve the surface of the course. James also became a proficient club maker using local wood from thc estate for the task. In addition he made gutta percha golf balls, and even found time to control the caddies who remembered him as a strict task master. His son, also James having started as a caddie, became the professional in succession to his father in 1902 at the early age of seventeen. Joseph was the President of the club and Arthur Read its Captain and the entrance fee was 2 guineas (£2.10), as was the Annual Subscription and by the end of the first year there were 100 members, subsequently rising to 125. A thatched building on the estate within 200 yards of the railway station served as the somewhat primitive changing room and clubhouse. Joseph, with his usual presence of mind had previously sold a parcel of the estate land opposite the station to the Friary Holroyd Brewery Company for the building of a hotel on the site, which was named the Ashley Park Hotel with Joseph's permission. The hotel duly opened for business in 1891 from which date it acted as the headquarters of the Club providing suitable refreshment for the players and accommodation for those who came from farther afield.

Joseph became very keen on the sport and was able to play off a handicap of five, as did Arthur Read. Joseph's second son, David appeared to be the most proficient of the family golfers playing off a handicap of three. It would appear that after leaving Ashley Park, David continued to enjoy the game and playing for some years following his move to Kirkcudbright. It should not be assumed that they were all expert golfers as the handicapping system at that time was far less rigorous than it is today. Nevertheless it shows that at least they attained a fair degree of proficiency at the sport. Teddie and Freddie were also known to have played and it is therefore likely that Totts and Arthur also did so. Louise was photographed wielding a golf club with some style, so it is possible that she at least played some golf, although a lady's attire at that time was somewhat inhibiting of

Ashley Park Hotel c.1900. *(Elmbridge Museum)*

Ashley Park Golf Club clubhouse. *(Elmbridge Museum)*

a free swing. In 1904 a cricket match, thought to have been an annual event, took place between the cricket and golf clubs.

The Ashley Park Golf Club flourished until 1905, by which time Joseph was finding that the cost of the expensive education of his many children coupled with the heavy maintenance costs of the house and estate had greatly depleted his financial resources. He came to the conclusion that it would be necessary to realise the value of a part of his estate by sale for development, and the obvious area was that nearest the station which was currently occupied by the golf course. Typically, before carrying the matter further, he considered it his duty to ensure that the club members were not inconvenienced more than necessary. Viscount Iveagh head of the Guinness family (later to become the Earl of Iveagh), had some years previously purchased the Burhill Estate in Walton on Thames, less than one mile from Ashley Park conditional on allowing the Dowager Duchess of Wellington to continue to live in the main house on the estate for the remainder of her life. As she had recently died Viscount Iveagh was approached by a syndicate, including both Joseph and Arthur Read, with a view to either purchasing or leasing the house and estate from the Guinness Family Trusts for use as a golf course and he proved sympathetic to the idea. Accordingly a committee was formed which included Joseph Sassoon, Arthur Read, The Hon Rupert Guinness and six others who negotiated a twenty eight year lease commencing at Christmas 1906, for Burhill House and the surrounding estate extending to approximately 300 acres for the formation of a golf club with two eighteen hole golf courses. The main house was to be used as the Clubhouse, a part of which was set up as a 'dormy house' providing overnight accommodation for the golfers travelling from afar. In the agreement there was an option for the club to purchase the freehold at a fixed price at any time during the tenancy, an opportunity which was not subsequently taken up. All the members of Ashley Park Golf Club were offered membership of the newly formed Burhill Golf Club and the continuity of golf in Walton on Thames was thus ensured. H.R.H. Prince Alexander of Teck was appointed President of the new club, and Arthur Read was elected as its first Captain when it opened for play in the following spring. Joseph and his sons David, Teddie and Freddie became members of the new Club.

At the final dinner marking the winding up of the Ashley Park Golf Club, held appropriately at the Ashley Park Hotel in May 1907, presentations were made to Arthur Read of a gold watch in recognition of his seventeen years as club captain, and to Joseph Sassoon of an illuminated address signed by all the members of the club, which bore the following words:

> "The members of Ashley Park Golf Club desire to express by this address their deep sense of gratitude to their President, Joseph Sassoon Sassoon J.P. for his unexampled generosity in allowing them the free use of his Park for golf during the last seventeen years and also having acted as their President for this long time, and for countless acts of kindness shown to them since the club was started in 1890."

As outlined above the Steer family had a major impact on the golfing scene in Ashley Park, but their connection with golf did not end there. In October 1905 a year before the club closed, no doubt having been tipped off, James Steer junior applied successfully for the assistant professional position at the Lytham & St. Annes Golf Club (before it acquired the Royal prefix), backed by an appropriately worded letter of commendation from the Secretary of the Ashley Park Golf Club. By 1908 he had taken up the appointment as the professional and club maker at Blackpool Golf Club, where he was in due course joined as fellow professionals by his brother Herbert and his son Gordon. The reason for the additional staff was that James had by then acquired a considerable reputation in the North West of England as a golf architect, designing and improving many courses there, having learned the craft from his father's early experience in Ashley Park. Despite these additional activities he nevertheless remained the senior professional at the club until it was requisitioned by the Air Ministry in 1940 when he became professional at Lytham Green Drive Golf Club until his retirement in 1950. James Steer senior, having been bitten by the golf bug, did not remain at Ashley Park for long after the demise of the golf club there. In 1908 he became professional at Harwich & Dovercourt Golf Club where he remained until his retirement in 1923. Two other sons of James senior also became golf professionals, Edward at Grange over Sands Golf Club and Joseph at Pendle Forest Golf Club and later at Burnley Golf Club. All were members of the Professional Golfers Association, quite a golfing dynasty all triggered by Joseph Sassoon's request to his gardener back in 1890!

The Sassoon family also became interested in the growing sport of lawn tennis and Joseph had a court laid out on the lawn close to the house. All the family were introduced to the game and as they grew up Joseph and Louise hosted many tennis parties there for their friends.

The family did not appear to take any real interest in shooting and accordingly the shooting rights over the estate were let out to a local resident, Walter Key who lived at Cypress Lodge from 1874 until his death in 1908. It is quite probable that Joseph's sons, given their military leanings, would have joined in the shooting parties on the estate in the later years.

Joseph did not show any great interest in taking part in the hunting scene but on many occasions acted as host to meetings which took place over Ashley Park and neighbouring estates. There is a reference in 1895 to the Surrey Staghounds cornering a buck on the estate at their meet. Both he, Louise and their sons and daughters participated in the social activities associated with the hunt and as an accomplished horseman he did occasionally participate in the chase although he did not enjoy the killing. On at least one occasion Louise was persuaded to join in, although she would not have claimed to be an expert horsewoman.

During the period from 1912 up to the Great War there was a revival both of paperchasing and drag hounds involving the following of a laid out trail on horseback over fences and ditches. The twins Freddie and Teddie who were skilled horsemen and who had taken advantage of the opportunity to participate in these sports at Eton, both showed keenness. Freddie showed an interest in polo, at which he played to a good standard. However Joseph stopped short of laying out a polo ground in the Park! He willingly allowed the estate to be used for various riding activities and on a number of occasions when the twins were joined by their fellow officers and by friends, but he was quick to point out in the press coverage that no cruelty to animals was involved, which views would have been much in the character of this gentle man.

The local athletic club were also permitted to use the Park for their sports days, at which Arthur Read took a keen interest and acted as their starter.

Chapter XII

The children fly the nest

The last decade of the nineteenth century marked the high point of the Sassoon family in Ashley Park, as it did for the Country as a whole. As has already been seen the house contained a lively family supported by a retinue of servants with Joseph at its head. He was aged 44 at the end of the decade, set in his ways, gradually retreating more into his shell, but still active on the sporting field and in particular on the golf course. Louise was not yet forty, she had borne seven children the youngest of whom were lively seven year old twins. She had taken a growing interest in village affairs and ran the large household with authority. The servants and her family had learned that it was best to keep out of the way when her temper was roused. She and Joseph had grown closer as the years progressed and photographs taken at that time bear evidence of their happiness together. The large retinue of indoor and outdoor servants looked after all their needs. The ever present Arthur Read was still coaching the younger generation. Totts the eldest son, had completed his first year at Eton and was enjoying his time there and his absence from Ashley Park in term time was barely noticed in the household. It was now clear that David would follow him there within the next year or so. Joseph's sister Rachel was flying high in London editing both *The Observer* and the *Sunday Times*. The only cloud on the horizon appeared to be the Boer War which had recently begun and was not going well but it was a long way away from Ashley Park. Jesse Trigg who had become the farm bailiff shortly after Joseph's marriage resided in the Bailiff's Cottage on the estate. His son, who bore the same Christian name, had recently started a milk round with Joseph's approval selling surplus milk from the farm. His enterprise flourished and grew into Trigg's Dairy with premises in the High Street alongside Church Walk which was the original pathway from Ashley Park House to the Parish Church. These premises exist to this day as a part of the Unigate Dairies plc. All appeared set for the next century in the unchanging life at Ashley Park. The happy family living in a large, well staffed house surrounded by a large estate including cricket ground and golf course looked after by an appropriate team of gardeners, greenkeepers and estate workers. Or was it?

In the census of 1891 Joseph had described himself as a "gentleman living on his means". As we have seen Joseph had never held any form of employment from which to earn an income, the estate land was of poor quality unsuited for commercial farming and the house was a source of ever more expensive maintenance. Joseph's sources of income were the interest earned on the inheritance

Vera Rosenburg with Joseph and Arthur 1908.

from his father's estate, rents from a small number of properties on the estate which were let, plus the proceeds from the occasional sale of plots for building on the periphery of the estate. In reality he was already living well beyond his means and his expenditure on education of seven children was about to add a further burden.

Horace Gunzburg had provided Louise with a capital base when she moved to Ashley Park but the investments were almost wholly in Russia and in particular in railway stocks there. Russia as a country was appearing less stable each year and the value of her capital was falling. Louise had been advised to diversify her investments by realising a part of her holdings and reinvesting the proceeds in stocks of her adopted country. Horace, however, strongly opposed this move, on patriotic grounds rather than on investment considerations, a stance which was to cost Louise dear.

By the middle of the next decade Arthur had followed his elder brothers to Eton but shortly thereafter David had left the school on account of ill health. He was showing an interest in painting whilst convalescing at home before enrolling at the Slade School of Art and Totts had moved on from Eton to his father's college at Oxford University. Joseph had by now regained ownership of the land south east of Ashley Road at a cost of £2,000, to finance which payment some further building plots on Station Avenue and in Ashley Park Road close to the Station had been sold. The twins Freddie and Teddie had started on their brief preparatory school education at Sandroyd school at Cobham before going on to Eton, and Hannah's health was causing concern. In term time the house at Ashley Park was no longer the playground of young children. It remained alive in the holidays when the family were often joined by relatives and friends. Among them was Louise's cousin Vera Rosenberg who hailed from Germany. She was attending a boarding school in England and began to spend her school holidays in Ashley Park. There she enjoyed happy times with the teen age children and more than twenty years later was to marry David. Louise with more time to spare entered further into the life of the village, becoming involved in events in connection with the St. Mary's Parish Church despite her Jewish faith. Rachel's husband Frederick Beer had died after his long illness, Rachel inheriting his fortune. Although this enabled her to live in comfort she had been certified insane and Joseph had been appointed by the Court to manage her affairs.

By this time Joseph had become all too well aware that he needed to take drastic steps to realise a significant portion of his assets to cover his growing expenditure. Within the next year he would be incurring the cost of university or public school fees for his five sons. In consultation with his land agents, Waterers, he reluctantly decided that it would be necessary progressively to sell a large part of his land for residential development. It was clear that area which lent itself most readily for such a sale was that land lying to the south east of Ashley Road, most of which was used as the golf course. Some building had already taken place around the borders but there remained nearly 100 acres on that part of the estate. During the next year the planning for the sale was advanced in some secrecy while negotiations, as has been seen, took place to lay out a replacement golf course in Walton, but outside the estate.

By February of 1907 boards had been erected on the site announcing the forthcoming sale and in June the following notice appeared in the *Surrey Herald* newspaper.

> "Freehold building estate of 80 acres recently occupied by Ashley Park Golf Club almost adjoining Walton and Hersham Station (the original name for Walton station prior to Hersham Station being built), near the river and Walton town with excellent shops. About one and a half miles from Burhill Golf Club and two miles from Brooklands motor track. The land lies high with gravel and sand subsoil, beautifully timbered, bounded by Ashley Road, Stompond Road and Hersham Road. The frontages to Station Avenue and parts of the

common and Stompond Road has already been developed and the remaining portion of the estate now being dealt with is fully ripe for immediate erection of superior class residences."

This part of the estate now became known as Ashley Wood Estate. Over the next few years Joseph extended Ashley Park Road and laid out new roads, Adelaide Road, Ashley Rise and Ashley Drive. He then sold off substantial individual plots on those roads on which high class houses were to be built. This development continued until the onset of the war brought such activity to a close for the duration.

The Walton Urban District Council had come into existance following the Local Government Act of 1894, encompassing Hersham and Oatlands as well as Walton. At the same time as these developments were taking place, Joseph offered to sell to the Council ten acres of land in Stompond Lane for use as a playing field at £500

Louise with her daughters, Mozelle and Missy c.1908.

per acre. On completion he was prepared to spend £500 of the proceeds in laying out the ground. This offer became the subject of heated debate in the Council being supported by most of the local tradesmen but the majority of the ratepayers were not in favour, so the offer was declined. More than twenty years were to pass before this same piece of land was acquired by the Council for such use but at a significantly greater cost. From the above it is not unreasonable to deduce that Joseph would have realised approximately £40,000 (nearly £2m at today's prices) when all the plots were sold. Even though the sale of plots proceeded slowly his immediate financial problems had been resolved.

At the end of the same year Totts, having obtained his BA degree at Oxford University, decided on a military career and was granted a university commission in the 6th (Inniskilling) Dragoon Guards. He thus became the first son to leave Ashley Park as he was required to live in the mess although he still spent his infrequent leaves at home. David was now studying art at the Slade school but continued for a short time to live at home whilst travelling to London for his studies.

Arthur, having followed Eton by Sandhurst was commissioned into 13th Hussars in 1909 and then left Ashley Park as did Hannah who for health reasons was residing in a Hindhead suffering from kidney problems from which she was to die there a year later.

In the same year David set himself up in a studio in Chelsea shortly before finishing at the Slade and accordingly spent less time at Ashley Park. He then studied further in Paris and travelled in Europe and in Algeria painting as he went.

The twins having left Eton for Woolwich and Sandhurst following which they were commissioned in 1912, Freddie into the Royal Artillery and Teddie into the 5th (Princess Charlotte of Wales) Dragoon Guards.

This left Mozelle as the only one of her generation to be still residing at Ashley Park which took on a more sombre appearance as the clouds of war gathered. The children despite leaving home never lost their affection for Ashley Park in which they so much enjoyed their childhood and a number of them featured Ashley in their name of their houses in the years to come. The Coronation celebration of 1911 was the last pre war occasion on which the grounds of Ashley Park were thrown open to the people of Walton, as 280 of their number dined and danced. Despite having left home not one of the sons was in a position to live either on the sale of paintings or their service pay but continued to depend on the financial support of their father, as did their sister. Cavalry regiments in the years of peace were a very expensive way of life for a young officer with high mess bills and the requirement to provide his own mounts so the sons continued to rely heavily on their father's generosity.

The rift which had existed between S.D.'s brother, Arthur and the Ashley Park Sassoons since the late 1850s was finally healed following a meeting between Joseph and his uncle in 1910. As a result of this meeting Arthur, who had no children, had become convinced that the Ashley Park children had a greater need of a share of his inheritance than his brother Reuben's children who had been nominated to benefit from his will by £100,000. Accordingly he set about changing his will, but before the revised document was signed he died unexpectedly. Reuben's children were thus to benefit rather than the Ashley Park children, but they made an ex gratia payment of £1,000 each to Joseph's children then living.

Joseph was by now suffering more severely from asthma and rarely left the house and even the golf course no longer held the same attraction for him. Louise had now become further involved in the village, taking an interest in a recently formed Walton Women's club by becoming their president. The club held a number of their meetings in the house and several functions in the Park. She no longer had the slim figure of her youth and photographs now showed a much more matronly appearance with grey hair but she had clearly continued to take good care of her appearance.

One hundred and fifty acres of farming land on the estate were now let to Jesse Trigg junior which provided Joseph with a small additional rental income. Jesse was also running a prosperous dairy business with premises in the High Street.

The population of Walton had now grown to nearly 6,000 with the proportion thereof being employed in agriculture having shown a marked decline. It could therefore no longer be correctly described as a village.

Chapter XIII

The Boys go to War

By the outbreak of the Great War four of the male members of the younger generation of the family were already serving as professional soldiers, only David of that generation, who had suffered health problems in his schooldays, remained outside the military regime. As soon as he heard of the declaration of war he returned home post haste from Algeria in an effort to enlist in the Army, only to be rejected as unfit for military service following his medical examination. Mozelle, not to be outdone by her brothers, soon undertook voluntary work with the Walton branch of the War Supply Depot.

The general impression at the start of hostilities was that with the advent of more advanced firepower the war would be of short duration. "Over by Christmas" was an expression frequently heard and there was a patriotic rush to the colours in order not to miss out on the action. It was expected that the role of the cavalry would be its traditional one of providing the mobility and in particular the rapid exploitation of breakthroughs by the infantry and thereby to cause chaos behind the enemy lines. The cavalry were the glamour boys of the Army and were generally expected to play their usual spectacular role. However, there were even at this early stage a few voices raised which expressed their concern that a European battlefield would not be like the wide open spaces of the Crimea, the Sudan or even South Africa, being the more recent of the occasions on which the cavalry had played a major part in the battles fought by the British army. Moreover they were worried that the advances in technology would make firepower much more effective and was therefore likely to tip the balance more strongly in favour of the defenders. These theories had not yet been put to the test in the heat of battle. It was with these uncertainties that the Sassoon boys embarked to test the lessons learned in their years of training as professional soldiers. The war has been fully analysed elsewhere so this narrative will concentrate on the roles of the individual members of the family.

Totts had been commissioned into the 6th (Inniskilling) Dragoon Guards at the beginning of 1907 and in the period before the war served with the regiment in England during initial training, then briefly in Egypt, followed by a tour of duty in India, the land of his father's birth, before the regiment returned to England. A fellow officer in the regiment was Captain Oates who obtained leave from the regiment to accompany Scott on the ill fated South Pole expedition and who was to die in such heroic circumstances. At first the peace time army service was easy going but as the threat of war with Germany grew more likely, the training became more intensive. Totts progressed in rank in the normal time scale and by 1913 he had been promoted to Captain.

When the war broke out he was immediately sent to France with what the Kaiser described as 'this contemptible little army', attached initially to 2nd (Queens Bays) Dragoon Guards. They were deployed in Belgium between the town of Mons and the French frontier and were involved in the battle there and in the subsequent 'retreat from Mons' at the end of August 1914, thereby earning the Mons Medal. Any visions of a short war (which had always assumed that Britain would be on the winning side) were swiftly dashed by the speed of the German advance. The fighting retreat continued until the front was stabilised between Ypres and the Belgian frontier.

Shortly thereafter, still in France he was attached to the Staff of the 1st Lancers, Indian Army in charge of British Army personnel. Early in 1915 he was pleased to be posted back to his own regiment. Alas his pleasure was short lived as the regiment was heavily involved in the abortive attempts by both sides to gain ground in the vicinity of Ypres in the course of which Totts was severely wounded in the lower jaw which took away a part of his face. These wounds were felt by the family to have contributed towards his early demise.He was returned to England for treatment followed by recuperation but to his chagrin his time as a fighting soldier was over and for the remainder of the war he was employed in the Directorate of Military Intelligence at the War Office in London.

As has been outlined David, with his weak heart, was declared medically unfit for military service but he was determined to join up and made further attempts to pass his medical, still without success. Still keen to contribute to the war effort he obtained work in a munitions factory. However after the heavy losses of military manpower in the Somme offensive of July 1916, the authorities were less stringent in applying the medical criteria and early in the next year David was finally accepted for military service. His low medical grading meant that he was not to be posted to a front line fighting regiment. Not for him the glamour of an officer in a smart cavalry regiment. He enlisted as a private in the newly formed Labour Corps (the forerunner of the Royal Pioneer Corps) joining the 138th Labour Company attached to the 2nd (Queens) Royal West Surrey Regiment with which he soon went to France and was firstly employed in the reconstruction of the damaged railway between Neuve-Eglise and Ploegsteert, just north of the Belgian frontier. Eton was not the usual breeding ground for privates in the Labour Corps who generally came from a more rugged background, some with experience of 'hard labour' courtesy of His Majesty. Their work as labourers and navvies was a far cry from the sheltered upbringing David had enjoyed, and it was to his great credit that he not only survived, but looked back on his experience with considerable pleasure. It also provided a good test for his weak heart, of which no more was heard! There followed a period as a dock labourer before he became secretary to the Staff Captain of his Company. He was granted a commission as a 2nd Lieutenant in the Labour Corps in November 1917 and continued to be heavily involved in the building of railways and roads, together with reconstructing trenches on the Western Front. During the last six months of the War he was following the advances closely, repairing and maintaining the roads to enable the forward troops to be kept supplied.

Arthur was commissioned into the 13th Hussars in 1909 and joined his regiment in India where it formed a part of the Meerut Cavalry Brigade which was in turn a part of the 2nd Indian Cavalry Division. By the outbreak of war he had been promoted to Lieutenant in the normal course, having had the opportunity to perfect his riding skills in the relaxed military regime using the wide open spaces of the sub continent. The regiment remained in India until November 1914 when it left for France, arriving at the front at the beginning of 1915. There then followed a period of frustration as the regiment was constantly on the move, taking up positions behind the lines ready to exploit the elusive breakthrough that never came. Much to his disgust they were also employed in the trenches at Neuve Chappelle and Theipval, so much for the mobility of the cavalry! Early in 1916 Sir Douglas Haig took over command of the British Army in France and Arthur was surprised to find that his cousin Philip, who had been his contemporary at Eton, and who had succeeded to both the baronetcy on his father's death and to his parliamentary seat, and was now Sir Philip Sassoon, the third Baronet and Member of Parliament for Hythe, and become Haig's private secretary. However Arthur's lot was soon to change as in June of that year his regiment was dispatched via Marseilles and Bombay (arriving appropriately at the Sassoon Dock) thence on to Basrah for service in Mesopotamia.

The background for this campaign was that Germany's allies, the Turkish army were threatening to advance through Baghdad and Basrah to India, thereby tying up British troops to counter the threat. When the Turks set up a base in Baghdad earlier in the war a British force had landed at the mouth of the Tigris and Euphrates rivers and advanced 300 miles upstream to Kut el Amera by the end of 1915 in order to put pressure on the enemy. This action was to lead to a disaster as the garrison there was surrounded by the Turks and starved into ignominious surrender early the next year when they became prisoners of war. As such they were ill treated with only a few malnourished survivors emerging from their their imprisonment. After prolonged discussion of the merits of further military operations in Mesopotamia by the Military High Command in London, it was decided to send a force of infantry and cavalry to avenge this reverse. Thus a new force was assembled from France and India which included the 13th Hussars, which would form a part of the 7th Indian Cavalry Brigade commanded by General Meade. The task set for this force was to retake Kut el Amera and to proceed onwards if possible to take Baghdad.

The hot summer months after the regiment's arrival were used for training and the force did not commence their advance until the beginning of November. Initially the advance was not opposed until it crossed the Shatt al Hai river in late December. Thereafter the mobility and aggressive tactics employed by the cavalry played a major role in the advance towards Kut el Amera which was retaken in February 1917. Without pausing the cavalry pressed quickly on and Baghdad was taken in early March. Thereafter consolidation took place before the summer heat demanded a further respite. Arthur, by now a temporary Captain and commanding a machine gun squadron played a significant part in these advances, being twice Mentioned in Despatches. Thus he spent the summer of 1917 in Baghdad, from which his great grandfather had fled nearly ninety years earlier. The campaign from his arrival in Bombay had thus followed in reverse the route of his great grandfather's flight from Baghdad.

The advance was resumed in the following November and continued through the winter months when it approached the strongly defended town of Tuz Khurmalti. Here Arthur's column of cavalry crossed the river Aq Su and knocked out a well sited machine Gun emplacement after a cavalry charge with drawn swords, capturing forty Turks and five Machine Guns. As a result of this action Arthur and two of his regimental colleagues were awarded the Military Crosses for conspicuous services rendered in connection with operations in Mesopotamia. He was promoted substantive Captain in April 1918 as Staff Captain of the 7th Cavalry Brigade with the temporary rank of Major following later. This was the last major offensive of the successful Mesopotamia campaign before the advent of summer again caused a break in hostilities. The now normal winter resumption of the offensive towards the end of October resulted in the capture of Mosul as the broken Turkish army now gave little resistance. This was the last battle in the campaign as shortly thereafter the armistice was declared. Some 400 miles had been covered since the landing in Basrah and the historic role of the cavalry, given open ground, had been vindicated, albeit against a less than well organised defence. In the honours awarded at the end of the War, Arthur was appointed an OBE and looked forward to the furtherence of his military career, but the star of the cavalry on horseback was by now well in the descendent.

Freddie having been commissioned into the Royal Field Artillery in July of 1912 had become the Royal Artillery fly weight boxing champion. At the outbreak of hostilities he had not yet been promoted and went straight away to France as a subaltern with the British Expeditionary Force and thus became the second 'Old contemptible' in the family. He was immediately involved in the front line across the Belgian frontier in the vicinity of Mons from whence they were driven by the initial German drive until the front was stabilised near Ypres, thereby following Totts by earning the Mons Medal. There followed a prolonged period of trench warfare in which the artillery played a prominent role. In this period Freddie was continually involved with little respite as his company moved further southwards towards the Somme. He was promoted Lieutenant in June of 1915 and to temporary Captain by the year end. As the winter wore on the front became a morass of mud and the morale of the troops was undermined by their living conditions in the trenches. The battle of the Somme designed to break the impasse, was launched on 1st July 1916 but proved a disaster with huge loss of life. Freddie was promoted to Captain (Acting Major) shortly thereafter. This was followed by further stalemate as the winter weather set in. For Freddie this meant the opportunity for a short leave in England to see his parents at Ashley Park. He was shocked how much they had aged and by the run down appearance of the Estate, without adequate manpower to maintain it to its former standard. His return to the front thereafter was short lived as he was severely wounded near Martinpuich, between Albert and Bapaume, which necessitated hospital treatment in England. It was not until August that he was declared fit for active service but he had put his leave to good by use becoming engaged to Margaret, the daughter of Frederick Franklin of Lancaster Gate, Bayswater. The family were well known to the Sassoons and Margaret was a great favourite with more than one of the Sassoon boys. Her elder sister, Dulcie, was at this time becoming close to Totts. Freddie was not to return to the Western Front but was attached to the Egyptian Army in the Middle East where he served as a Bimbashi (Captain) from August of 1917 until the end of the War.

Teddie was commissioned into the 5th (Princess of Wales's) Dragoon Guards in February 1912 in which regiment he served initially in England and then in Ireland until the outbreak of War when he was promoted Lieutenant as the regiment went to France as part of Allenby's 1st Cavalry Division. Whilst there, frustrated by the lack of opportunity for the cavalry, he volunteered to be an Observer attached to the newly formed Royal Naval Air Service. In this hazardous task he served in France until the summer of 1915 when he was posted to a similar role with the Eastern Mediterranean Squadron in support of the ill fated Gallipoli landings. He had served there only three months when the decision was made to evacuate the peninsula. A number of the New Zealand troops who had served alongside him in that campaign and had been severely wounded there, found themselves shipped to the wartime hospital for New Zealand troops at Mount Felix, adjoining the Ashley

Park Estate. On returning to England, Teddie was attached to the Berkshire Yeomanry as a part of the Reserve Cavalry Regiment, remaining in England until August 1917 when they formed part of the Egyptian Expeditionary Forces in Palestine once again coming under the command of Allenby. The campaign culminated in the taking of Jerusalem at the end of that year. At the termination of the War he was serving there with the Army of Occupation.

It can thus be seen that all five male 'Ashley Park Sassoons' played their full part in the War, seeing service at the front in all the major campaigns, in Belgium, France, Gallipoli, Mesopotamia Egypt and Palestine. It is remarkable, bearing in mind the level of casualties experienced in this War, that all five of the sons survived the hostilities although two of their number were badly wounded. As professional soldiers (with the exception of David) following the cessation of hostilities they might have looked forward to the blossoming of their careers but with the cavalry regiments suffering drastically from the post war amalgamations and cuts, their future proved to be distinctly bleak. Within five years not one of them remained in the service. Thus for them the only career for which they had been trained was closed, leaving them as beneficiaries of a small capital sum from their father's estate which was clearly insufficient to provide an income on which they could live in any comfort. Thus the outlook was bleak in the extreme.

Chapter XIV

The Home Front 1914-18

As Joseph and Louise bade farewell to their male offspring at the beginning of the war their feelings must have been mixed. Whilst they would have expected their sons' careers as professional soldiers to be advanced, they were well aware of the inherent dangers which faced them. Joseph was aged 58 and his health, plagued by asthma and not helped by his smoking, was showing signs of deterioration but Louise aged 52 appeared to gain in strength as his health declined. However, photographs taken at this time showed that she had begun to put on considerable weight and the early symptoms of her future thyroid problems had become visible. Like most of their contemporaries Joseph and Louise initially expected that the War would be of short duration and that life at home would continue little changed.

The first casualties of the war were the innocent Belgian civilians driven from their homes as the initial German onslaught, aimed at France, by-passed the French defensive wall by driving through neutral Belgium. As the flow of Belgian refugees grew rapidly, their plight attracted sympathy throughout England, as attempts were made to find them what was expected to be temporary accommodation here. Within ten days of the outbreak of hostilities a local relief committee to deal with their homelessness was formed in Walton under the chairmanship of Sir Godfrey Langdon on which Joseph served. He immediately put one of his properties, Woodcote in Ashley Road at the junction with Oatlands Chase at their disposal and 26 refugees were soon housed there. Shortly thereafter another of his properties, Ashley Warren in Oatlands Chase opposite Oatlands Avenue was also to be made available for the same cause as the number of refugees grew.

In 1899 Cecil Hepworth had purchased a house in Hurst Grove, off Bridge Street which adjoined Ashley Park. Within a year he had started to produce films there and by 1905 had built his first film studio in the grounds of his house. Shortly before the end of 1914 he agreed to provide a Drill Hall and Rifle Range for the local Volunteer Training Corps. In seeking a site for its erection he approached Joseph to ascertain whether he would agree to site it in Ashley Park. Joseph responded straight away by making available a site on which the building was duly erected. Despite being blessed with these assets the local corps was never a success and was suspended before the end of 1915. This temporary building was demolished at the end of the War.

Early in that year Joseph and Louise were distressed to learn that Totts had been severely wounded in the lower jaw but pleased that he would be returning to

Totts at Ashley Park in uniform.

England for hospital treatment. Their concern over his injuries was tinged with relief that their eldest son and the heir to Ashley Park would recover but was unlikely to return to the front, although he would be able to continue his military career at the War Office in London. Alas the wound was to have a much more serious effect on his future life.

Joseph also threw open the grounds of Ashley Park to the military for training and the Honourable Artillery Company and the Westminster Volunteer Rifles made extensive use of it for camps and military exercises including trench digging. A fete was held in Ashley Park during the summer in aid of the Red Cross to which Louise brought a large party. The cricket ground was the venue, as it was no longer being used for the game as all civilian sporting activities had been suspended for the duration, although the armed services continued to play rugby and football in the Park.

In July New Zealanders living in England raised funds to set up a New Zealand War Contingent Hospital and sought appropriate premises. Mount Felix, a large property on the opposite side of Bridge Street from Ashley Park was found and requisitioned to provide treatment for war wounded New Zealanders. Initially it treated less than a hundred patients, who were immediately allowed access to Ashley Park for exercise. Following the Gallipoli landings in 1915, which involved a high proportion of ANZAC troops and their ensuing severe casualties, the hospital was officially taken over by the New Zealand medical authorities who rapidly increased the number of beds to five hundred. By 1917 the number of wounded New Zealand soldiers at the Mount Felix hospital had grown to a peak of over 1,500 soldiers. This expansion required the building of additional hutted premises in the grounds. In addition an annex was established when the Oatlands Park House was requisitioned. In July of that year Joseph and Cecil Hepworth, who had both supported the hospital throughout its existence joined forces to entertain a gathering of two hundred wounded soldiers and their nurses from Mount Felix in Ashley Park. Alma Taylor, a film actress from the Hepworth Studios was the star attraction, signing photographs for the wounded men and acting as hostess.

Early in 1916 as the allied shipping losses mounted the government were concerned that an invasion of England could not be ruled out and all residents of Walton and Oatlands were advised that in the event of such an occurrence they were to assemble in Ashley Park and there to await further instructions. By the middle of the year more recruits were needed and conscription was introduced. Joseph and Louise, like many other land owners, were therefore finding it difficult to retain sufficient staff to maintain their previous lifestyle, and gradually more rooms in the house were shut down. The estate was also becoming increasingly overgrown through the lack of maintenance.

By the beginning of 1917 the government was trying to identify uncultivated land which could be used to increase the food supply as the import of foodstuffs became more difficult as German U-boats harried the merchant ships. The Royal Navy fought a desperate battle to keep the shipping lanes open. Joseph's response was typical in offering land at Ashley Park, but this offer was declined "as the land was of poor quality and no good for cultivation without fertilisers." At this time Joseph and Louise were further disturbed to hear that Freddie had been wounded in France and was being shipped home to hospital. As he convalesced they were cheered by his sparkle, which helped lift the gloom which had descended on the remaining inhabitants of Ashley Park. There were two happy family occasions to lift their spirits. Firstly Freddie's engagement to Margaret Franklin which was welcomed by both families, followed shortly thereafter by Mozelle's engagement to Lieutenant Edgar Spielman whose family were well known to the Sassoons, but who was neither liked nor fully accepted by Joseph and his sons.

Edgar had been admitted as a broking member of the Stock Exchange in 1908 and on the outbreak of war joined the Honourable Artillery Company as a private soldier and served in Belgium and France before being invalided home at the beginning of 1915. After recuperation in March of that year he obtained a commission in the 25th (Cyclist) Battalion, the London Regiment, in which he served on the coastal defences in Norfolk. By the time of his engagement he had transferred to the 31st (Wales) Battalion, the Middlesex Regiment involved in aerodrome reconstruction in Essex and Kent.

Shortly thereafter Joseph and Louise reluctantly came to the conclusion that they had no practical alternative but to give up the unequal struggle to maintain Ashley Park, as the War continued to drag on with little sign of an early end. Joseph was worn down by his efforts to maintain the house and estate, by the attacks of asthma which were causing a strain to his heart, and the constant worry as to the well-being of his sons. He had further retreated into his study, whilst Louise had become increasingly concerned about the state of his health. With the utmost sadness they came to the conclusion that they had to vacate Ashley Park at least for the duration of the War and move into more manageable accommodation locally. Their land agents drew their attention to the possibility of taking a lease at £100 a year on a property named Springfield in Rydens Road which was only a quarter of a mile from the edge of the Park. Rydens Road was lined with substantial houses on the north side and farm land between it and the railway to the south. Springfield was a large but somewhat depressing detached house set in over an acre of ground on the north side of the road, having eight bedrooms, two drawing rooms three reception rooms together with stabling and a garage. Today the house still stands but having been subsequently sub-divided into several flats and in the grounds three further houses have been built. It is clear that even in its hey day, this must have been a devastating move for the unhappy couple. Into this house moved Joseph, Louise and the ever present Arthur Read, together with a reduced number of servants. Joseph had only known and loved one property since he was seven years old, and he was never to recover from the parting, even though he still owned the Ashley Park House and estate and hoped to return there following the cessation of hostilities. In the meantime Ashley Park House, and the garden, was offered to let on a short lease, initially without success.

A few months before they left Ashley Park, Joseph was visited by his cousin David Solomon Sassoon, the son of S.D.'s brother Solomon, who created the Sassoon library of over one thousand books and manuscripts, both Oriental and Western, considered to be the greatest privately held collection of Hebrew manuscripts. He shared Joseph's scholarly interests, took a great interest in the library at Ashley Park and held him in high regard. Following his visit David described, in a letter to his mother, the sad visit he made to the now almost deserted Ashley Park. He was later to write, following Joseph's death that "He was the **only** member of the family who did not talk of **money** and had a liking for knowledge, education and literature. I was much struck, when he came to see my books, with his knowledge of Jewish things." High praise indeed coming from such an eminent bibliophile.

In November 1917 Mozelle and Edgar were married at the West End Synagogue in Bayswater. Mozelle wore a white satin dress covered by spotted net, trimmed with old Brussels appliqué lace with a full court train, a gift from her paternal grandmother, Flora. Joseph was well enough to give away his daughter and was cheered by the event. Totts acted as best man and the two bridesmaids were Dulcie Franklin, herself later to marry Totts, and a cousin Muriel Sassoon, the daughter of S.D.'s recently deceased younger brother Frederick, who at the time of his death was the Chairman of David Sassoon & Sons.

The year 1918 dawned without any respite in the War although the recent entry of the United States giving promise of significant additional manpower and munitions had given hope for the future. However, 2nd January brought great sadness with the not entirely unexpected death of Joseph, who suffered a heart attack two days after his 62nd birthday. The ever faithful Arthur Read was present at his passing and signed his death certificate along with Norman Rushworth, the family doctor and friend of many years standing. As the certificate shows, the death occurred at Ashley Park, whence Joseph had chosen to be moved from Springfield in anticipation of his demise. To those close to him it was clear that his health had been noticeably in decline since the outbreak of War, nevertheless his passing came as a great shock to the family, particularly his sons fighting overseas who had seen little of him in recent years. The coffin passed between the depleted lines of indoor and outdoor servants and through the main door of Ashley Park House which he had first entered fifty five years before.The funeral took place at the Spanish & Portuguese Burial Ground in Mile End where his grandmother Flora, his father, his brothers Frederick and Alfred and his daughter, Hannah were interred. As his death came without warning, none of his sons who were serving abroad, were able to attend the funeral. Only Totts who was at the War Office in London was able to be present at the ceremony, a sad end for the man who had been at the centre of Ashley Park life for over half a century.

In his will made twenty three years before his death Joseph appointed as his Executors Sir Edward Sassoon, the second baronet (son of Sir Albert), Baron Vladimir Gunzburg (Louise's brother) and the family solicitor B.F. Hawksley. In the event Sir Edward had predeceased Joseph by some years and so was unable to act. By the will, Louise was to enjoy the land and buildings for her life together with the income from the estate all of which would on her death pass to Totts absolutely. Joseph's estate at his death amounted to only £43,000 (equivalent to over £850,000 at today's values), a surprisingly low figure, most of which was tied up in the property which was mortgaged, so there was now no possibility of Louise returning to Ashley Park, even if she had wanted to do so which was unlikely.

As a consequence there were sales of furniture, paintings, books, a grand piano and a full sized billiard table from the house at both Christie's and Waterers which realised in excess of £3,500.

Towards the end of the same year the War came to its successful conclusion, which coincided with the letting of Ashley Park House at a rent of £600 a year to a Mrs Graeme Gibson, who was to prove to be a continual headache throughout her tenancy.

Chapter XV

Running down

There was immediate speculation, following Joseph's death and the departure of the family from Ashley Park, that the estate was likely to be put on the market. At this time the Walton Urban District Council were drawing up their first Town Plan and approached Totts as to the possibility of 'opening up' the Estate. Clearly this could not be done in Louise's lifetime without her support for any proposal but he did not rule out the possibility out at some date in the future. From this time the Council made their plans for that event.

The year 1919 started with two joyous events following closely on one another. Firstly Totts' marriage to Dulcie Franklin, the daughter of the recently deceased Frederick Franklin, at the home of Dulcie's maternal grandfather, Philip Magnus to be followed shortly by the marriage of Freddie to Dulcie's sister Margaret (Lucy), held at the same residence. The ceremonies were somewhat subdued as the Franklin family were in mourning for the girls' father. It was at that time by no means uncommon amongst the wealthy Jewish families to hold the wedding ceremonies in the home.

These events were followed in April by the death of Flora, S.D.'s widow in Hove, at the age of 82. She had been widowed for over fifty years and had outlived all her children with the exception of Rachel, who had a cocooned existence in Tunbridge Wells. As Flora had grown older, her visits to Ashley Park had become less frequent as her grandchildren had grown up, and she did not enjoy making the journey. In addition she and Louise had never enjoyed a close relationship. She had not altered her will to reflect Joseph's death a year earlier and as he was named as the sole beneficiary of her estate of £8,068 (£158,000 at today's values), it would pass in equal shares to her Ashley Park grandchildren.

In the middle of the year there was further cause for celebration as Arthur was appointed an Officer of the newly formed Order of the British Empire (Military Division) in recognition of his splendid war record. His Army career continued to flourish with his appointment as Deputy Assistant Adjutant General of the 18th Infantry Division in the Middle East.

The summer saw several events to celebrate the coming of peace and Louise readily agreed that the grounds of Ashley Park once again be the focal point of the festivities in Walton on Thames. The diminishing number of New Zealand soldiers who had remained in Mount Felix until the hospital was finally closed down, provided the decorations for the entrance to the Park and a crowd estimated at more than 2,000 was present. The event was somewhat marred by the fact that

Louise, who was due to perform the opening ceremony, was not able to do so as she was unwell. Prior to this Louise's health, apart from some problems of late with her thyroid glands, had given little cause for wider concern. This was the last of many momentous occasions to have been celebrated in the Park during the period of the Sassoon ownership. Shortly after this event the hospital was closed down, having provided hospital care for a staggering 27,000 servicemen in its short life. Twenty one New Zealanders who died of their wounds are buried in the cemetery adjoining St. Mary's Parish Church in Walton and a memorial was erected in their memory. Each year up to this day a service has been held in the church on Anzac Day, attended by the New Zealand High Commissioner, to remember those who died in this town, so far from their home.

In October it was decided to sell Ashley Cottage and its surrounding 9 acres of land which had been occupied for some years by F.R. Spofforth, the 'demon' Australian bowler of the 1880s who had recently moved to Surbiton. In the event no purchaser was found so the sale did not take place, and it was later to be included with other properties when they were auctioned more than three years later.

In March 1920 Totts and Dulcie celebrated the birth of a daughter who they named Penelope Louise. That summer saw the re-formation of Walton Cricket Club who were as yet unable to return to their ground at Ashley Park as it had not

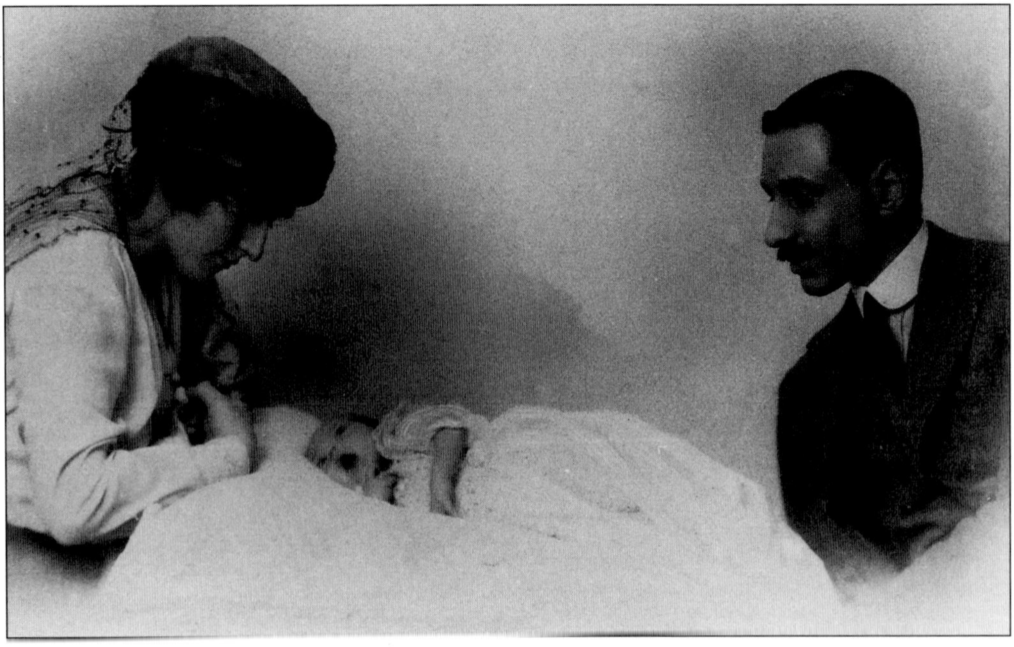

Dulcie, Penelope and Totts 1920.

been restored to its former condition after lying fallow during the War. Arthur Read who had looked after the club's equipment with great care since stumps were drawn at the close of the 1914 season, formally handed it over to the re-formed club who were to play their matches for this year at a ground at Mount Felix. This had been the estate once owned by the Earl of Tankerville, who employed as a gardener and cricket professional 'Lumpy' Stevens, the bowler credited as being the cause of the introduction of the third stump as his accuracy was such that the ball frequently passed between the two stumps then used.

In the same year the District Council entered into discussions with Totts and Hepworth & Co. with a view to the building of a new road from the junction of Church Street and the High Street to Walton Bridge and the development of the surrounding land. The land was subsequently acquired but in the event this road was not to be completed until the 1960s due to a lack of finance.

March 1921 saw the death of Louise at Springfield in her sixtieth year. She had never fully recovered from Joseph's death and her health had steadily deteriorated as goitre was diagnosed and she tired quickly. She had continued to receive medical attention to which she did not respond, and it was now clear that she had not long to live, so that her death when it came was no surprise to her children. She was interred next to her late husband in the Spanish & Portuguese Burial Ground in Mile End. She left her estate which was valued at over £11,000 (over £250,000 at today's values) to Totts.

Unfortunately by this time his own health was giving increasing cause for concern as cancer had been diagnosed arising from his war wounds and he could not expect a long life. It was now clear that a further death and another payment of Estate Duty would mean that the family would be in no position to move back into Ashley Park.

Meanwhile the tenants of Ashley Park were causing a nuisance in many respects. Advertising hoardings appeared without permission on the walls of the estate fronting the High Street and Oatlands Drive causing outrage in the village and Totts had to resort to legal proceedings to obtain their removal. Reports relayed to Totts suggested that forty rooms in the house were being used as an unregistered hospital for profit to the benefit of Mrs Graeme Gibson and Miss Murray housing mental patients, contrary to the terms of the lease. Villagers were afraid to cross the park because of the activities of the inmates. A fire occurred in the outbuildings but prompt action by the local fire brigade prevented it spreading to the main house. The tenant's husband, Mr Graeme Gibson appeared on a number of charges of failing to pay maintenance to a young girl who had bourne his child and who was living at the Sailors Home, an inn in St. Mary's Road, Oatlands. He was also summoned for arrears of rates. Miss Murray who was a relative of the lessee had failed to make the required payments on furniture acquired for the hospital on hire purchase and was sued for their recovery. Clearly this unsatisfactory state of affairs could not continue and two years later Miss Murray was declared bankrupt and the lease fell in thus ending a sorry episode in the history of the estate.

By the middle of the year Totts had been forced reluctantly to the conclusion that the only sensible course was to dispose of the estate. No doubt aware of the disputes over inheritance which occurred in the family in the previous generation, he discussed the situation with his brothers who concurred. It was also agreed between them that the proceeds of the sale should be shared equally amongst them, after making provision for Dulcie. Accordingly he consulted his land agents, Waterers, and plans were set in train to prepare for the sale of the estate by auction.

The year 1922 commenced with the happy news of the engagement of Arthur to Doris the younger daughter of Mr and Mrs Elias Meyer of Grosvenor Square, whose family originated in Calcutta having similar Indian roots to those of the Sassoon family. They were married later in the year alas without the presence of his eldest brother whose health had gone downhill rapidly.

Totts passed away after a painful illness on 7th February of that year. Although the cause of death was cancer of the pharynx for which he had been receiving treatment for eighteen months, the family were convinced that his death followed directly from his war wounds which were to the lower jaw. Totts broke from his family tradition by being buried at the Jewish cemetery in Golders Green. His wife Dulcie and his brother, David, together with Totts' solicitor were named as executors of the will made shortly after his mother's death. His estate was valued at approximately £225,000 (£5.2m at today's prices). This represented a substantial increase over the amounts left to him by Joseph and Louise and would appear to take account of the enhanced development value of the Ashley Park Estate following the decision to sell. In his will Totts left to Dulcie, whose family had substantial wealth, £2,000 free of tax plus chattels, horses and cars. To Mozelle he left the income for life from £6,000 and thereafter to her issue. An annuity of £300 a year was left to Arthur Read who still lived in Walton but was now frail and two small annuities to long-standing servants of the family. The balance was left one third to Dulcie in trust with income for life or until remarriage. On either of these events capital and income would pass to their children. Trustees were authorised to pay up to £5,000 for the purchase of a house for Dulcie. The remaining two thirds to be divided equally between the four surviving brothers. As their shares would exceed £20,000 each the will provided for a further £5,000 to be added to Dulcie's share. She also inherited Joseph's library which had passed to Totts and over the succeeding years she disposed of it piecemeal rather than as a whole, much to the chagrin of Sir Isiah Berlin, whose wife was a niece of Louise and who would have been interested in purchasing it in its entirety.

In the September of that year Arthur Read died at the Walton Cottage Hospital in his seventy second year having come to Ashley Park over fifty years earlier. The funeral was held in Walton Parish Church and he was buried in the churchyard there, where the headstone can still be seen. The remaining members of the Sassoon family were in attendance as were representatives of many sporting organisations of the town and its surroundings. Thus within a four year period after the family's departure from the Estate five of the persons most closely identified with Ashley Park had died, namely Flora, Joseph, Louise, Totts and Arthur Read. It is difficult to dismiss the feeling that their distress at the parting from a much loved home contributed at least in some measure to those deaths. The younger generation would now go their own ways but with a financial cushion against the immediate need to earn a living.

In April of 1923 the estate was put up for auction. The mansion and approximately 170 acres were to be sold as one lot and six houses in Oatlands Drive, Station Avenue and Oatlands Chase were to be sold separately. In addition there were 39 acres of building land being the remainder of the land lying east of Ashley Road which had only been partially developed. The auction was due to

Long Gallery in the empty house 1923. *(Elmbridge Museum)*

take place appropriately at the Ashley Park Hotel on 19th April but on the day of the sale it was announced that the entire Estate had been acquired by Mr A.O. Davis, a property developer, living at Heatherdene, Oatlands for an undisclosed price but, that the auction would take place as planned. An amendment was made to exclude from the sale all tenants fixtures, tapestry paper and the picture over the fireplace in the Chinese room in the mansion. Thus the Sassoon connection with Ashley Park ceased, no doubt with the purchaser obtaining a bargain at the expense of the family.

The Walton U.D.C. in a subsequent deal with Mr Davis obtained the land it coveted in order to widen the High Street. Later after a Mr Brassey Taylor had acquired the undeveloped parts of the estate, the Council acquired from him further land to build the new by pass from the northern end of Ashley Road to the bridge to be named New Zealand Avenue, to commemorate the wartime New Zealand military hospital at Mount Felix. However, due to the recession it was the year 1935 before the finance was available for it to be built. The Council also acquired the land south west of the by pass, to be kept as open space, which land included the old cricket ground which was returned to its former use. A new Town Hall for the Walton and Weybridge Urban District Council was built on part of the acquired land, and opened by H.R.H. Princess Margaret in 1966. The creation of the Elmbridge Borough Council to replace the U.D.C. in 1974 rendered the Town Hall inadequate for the Council's needs, and it was demolished in 1991, to be replaced by a Homebase store. The Council also purchased the land for a recreation ground in Stompond Lane which they had been offered and had declined many years earlier. The remainder of the land which formed the Ashley Park Estate was developed for housing, however, such development proceeded slowly, and was not completed until after the Second World War.

The Mansion House was repossessed from the defaulting tenants by Mr Davis and remained empty for two years during which time all the fixtures were sold, including the ornate staircase which was dismantled and taken to America. Following his acquisition of the estate Mr Davis managed to obtain possession of the various properties let to estate workers and retired workers at nominal rents, which properties were subsequently sold at full market value. This underlined the view that Joseph had been a most sympathetic and generous employer, but in later years he could ill afford such generosity.

Chapter XVI

What became of the survivors

David, the eldest of the remaining sons of his generation, like his father was of a retiring nature but more sociable and the friends which he made he kept for many years. In appearance he was not typical of an artist, looking more like a rather studious solicitor. He was demobilised from the army shortly after the war ended and became a Special Constable for a time before returned to his painting specialising in water-colours, some of which are in the possession of the Imperial War Museum. He now rented a property in Jubilee Place, Chelsea which his grandmother, Flora furnished for him and rejoined his pre-war friends in the artistic community there. His continuing friendship with Augustus John led to him being appointed guardian to Joy Hicks, Augustus' daughter by his long time model Jacintha. In this period he also became friendly with Joseph Simpson who had been an official war artist with the Royal Flying Corps in France. Through his influence David was first introduced to Kirkcudbright, which sported a thriving artist colony, and to the Galloway landscapes both of which were to feature large in his future life. His paintings were not however typical of the Kirkcudbright school being influenced more by the French and German expressionism.

In 1922 he exhibited his works in Paris but there is no record of the sales which ensued. Shortly thereafter he met again Vera Rosenburg, Louise's cousin, who as a schoolgirl had been a favourite of more than one of the Sassoon boys in the summer holidays which she often spent at Ashley Park in the pre war days. She had matured as a free spirit, being brought up on a loose rein in Berlin. She was the first of her sex to negotiate the Cresta toboggan run and learned to fly at an early age. *
The relationship grew and they became engaged.

It was to Berlin that David went for their wedding in 1926. Whilst there he found time to exhibit 46 of his paintings, arranged through Vera's family at the Paul Graupe Gallery. It is not known how many sales were made, indeed although he exhibited from time to time and painted throughout his life he did so for pleasure rather than profit and sold few of his pictures as he was too self effacing about his work, preferring to give them away to his friends. It must have been a great shock for the worldly-wise Vera to be taken to set up home in Kirkcudbright in the somewhat remote backwater of the South West of Scotland, but she enjoyed the bohemian atmosphere which pervaded the artist colony which flourished there. David's modest inheritance was further boosted following the death of his Aunt Rachel, and the low cost of living in Kirkcudbright enabled him to live comfortably without the necessity of earning a living: like father, like son! A year

* No. She was nearly the first woman to fly (with the Wright brothers) but sadly wind prevented the flight on that day.

David with one of his works 1926.

later their first son, named Joseph after his grandfather, was born, to be followed three and a half years later by a second son, Gabriel Jacques, named after his maternal grandfather.

During the Second World War he served as a Special Constable. Later he and Vera moved to London where they lived at the White House in Regents Park until Vera's death following a fall there in 1973. David then returned to Kirkcudbright where his son, Joseph and his grandchildren lived. He lived on for a further five years, dying on the last day of 1978, aged 90. Not at all bad for a sickly child forced to leave Eton with a weak heart! In a posthumous exhibition held in Kirkcudbright in 1979 there were shown fifty water colours and twelve oil paintings at prices which ranged from £130 to £1,500.

Arthur, having had a 'good war' soldiered on until 1922 having risen to be Deputy Assistant Adjutant General of the 18th Infantry Division. By that time it was clear that the days of the mounted cavalry had passed for good, to be replaced by tanks and armoured cars. Regimental amalgamations coupled with reductions of manpower were now the order of the day. Like so many of his contemporaries he was trained for no other occupation and like his brothers he showed no inclination for the routine of a business career. He did however respond to the call of duty and served briefly as a Special Constable during the short lived General Strike of 1926 and again during World War II. In 1922 as intimated he had married Doris Meyer who came from a very wealthy family, which provided a cushion for his lack of income. They produced three children, Beatrice, Hugh and Cynthia but the marriage ended in divorce at the end of the Second World War. In 1948 he was married for the second time to Yvonne Massay who claimed descent from Napoleon's Marechal Murat. They lived in West Africa but this marriage also ended in divorce eight years later. Arthur died penniless in 1960 a sad end for one whose early military career had promised so much.

After Edgar Spielman was demobilised he and Mozelle set up home together in Sevenoaks, from whence Edgar travelled to London to resume his career as a stockbroker. Soon Mozelle became pregnant but the daughter was still born. The distraught Mozelle called her Hannah after her deceased sister but sadly was unable to have further children. To make up for the disappointment they later adopted two children, Nancy and Brian. Edgar's stockbroking career which had been interrupted by the war, thrived once more and he became a partner in the stockbroking firm of I. Albury & Co. specialising in gold shares which together with Mozelle's inheritance from her Aunt Rachel enabled the couple to live in some style. Finding him somewhat stuffy and lacking in character, Mozelle's brothers never took to him. Had they known then that he was carrying on a long standing liaison with a lady from Worthing they may well have considered him to be a more interesting personality than they did. The outbreak of the Second World War saw a major setback in the couple's fortunes as the market in gold shares collapsed, but their fortunes recovered when peace returned and they lived out their life in comfort, moving within the Sevenoaks area once again, naming their house there Ashley. Mozelle lived on in until her death in 1973, aged 83, Edgar outliving her by five years.

Following his marriage to Margaret (Lucy) Franklin, Freddie and she set up their home in Barnes, in a house which he named Ashley Cottage. He was disgusted by his treatment by the army which placed him on temporary half pay at the beginning of 1920, before discharging him on health grounds as a result of his war wounds in August of that year, at a time when he considered that he had made a full recovery. They had three sons, Humphrey, Richard (killed in World War II) and John. Freddie was the exception of the Ashley Park Sassoons, working in the city for Gray Dawes on the shipping side of the business but did not rise to the top echelon of the firm. Lucy died in 1952 following which Frederick moved to Milton Abbas where he died in the 1970s.

Teddie was the extrovert of the family who radiated bonhomie in any company. After the War he was able to show sufficient wealth and became an underwriting name at Lloyds which at that time was enjoying a period of considerable prosperity, in which he would have shared. He remained fancy free until 1929 when at the age of 37 he married Minnie Heseltine (nee Channing) who was six years his senior. The wedding took place in the south of France and they continued to enjoy the Riviera climate with the exception of the years of World War II when he acted as press censor in Glasgow. They returned to France after the war where they lived on until her death there in 1954. He outlived Minnie by only a year.

In 1943 Totts' daughter, Penelope married Serge Gunzburg, a grandson of Horace, thus she was the third generation of the Ashley Park Sassoons to inter marry with three members of the Gunzburg family, Joseph having married Louise who was the daughter of Horace, followed by David's marriage to Horace's niece Vera Rosenburg and later Penelope to his grandson Serge. These marriages coupled with Totts and Freddie both marrying the Franklin sisters underlines the close knit group of prominent Jewish families who mixed with each other in a tight social circle.

In many respects, it can be seen that the rise and fall of the Ashley Park Sassoons reflected the path followed by the country in the same period. The last half of the Victorian era coincided with a long period of prosperity, similar to that enjoyed by the Sassoons. The reverses of the Boer War were the first signs of that period coming to an end which was matched by the early signs of financial strain felt by the Sassoons. The Great War, despite the final result, cost the country dear, both financially and in the loss of manpower. The same pattern hit the Sassoons who were no longer able to employ sufficient staff to maintain Ashley Park to its former standard. Although all five of their officer sons survived the war, within five years of its cessation four family members, spanning three generations met their death. With their passing and the introduction of death duties the Sassoons were unable to hold onto their Estate, a situation forced on many prominent families in the post war period.

The Sassoons of Ashley Park
Abbreviated Family Tree

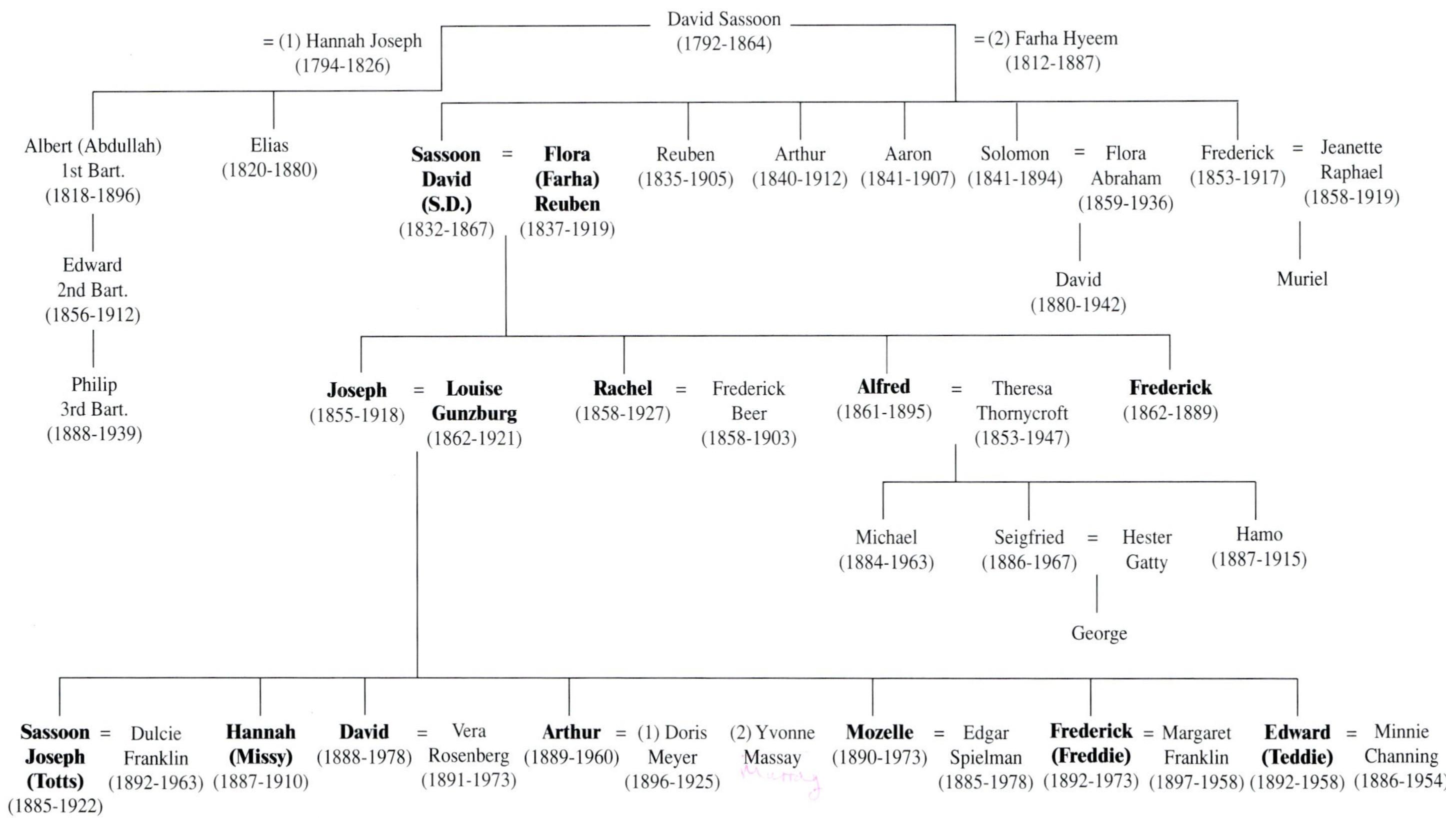

Note: **Bold type** = sometime resident at Ashley Park.

Index

Adam, Robert, 50
Adelaide Crescent, Hove, 38
Adelaide Road, Walton, 84
Albany, Duchess of, 67
Albert, Belgium, 91
Alexander III, Tsar, 37
Alexander of Teck, Prince, 79
Alexandra, Princess of Wales (later Queen Alexandra), 28, 52
Allenby, General (later Lord Allenby), 91, 92
Alma-Tadema, Sir Lawrence, 75
American Civil War, 25, 27, 28
ANZAC, 95
Anzac Day, 100
Aq Su, River, 90
Artom, Rev. Dr., 28
Ashley Cottage, 19, 22, 25, 52, 58, 75, 100
Ashley Cottage, Barnes, 107
Ashley House, Sevenoaks, 107
Ashley Drive, 84
Ashley Park Golf Club, 31, 76-79, 83
Ashley Park Hotel, 67, 77, 79, 104
Ashley Park Road, 34, 83, 84
Ashley Rise, 84
Ashley Road, 24, 34, 67, 75, 82, 83, 93, 103, 104
Ashley Terrace (see Churchfield Road)
Ashley Warren, 93
Ashley Wood Estate, 84

Baghdad, 9, 11, 12, 15, 26, 29, 89, 90
Balaclava, 68
Bapaume, 91
Basrah, 10, 89, 90
Beckett, Arthur, 58
Beeches, Oatlands Drive, 74
Beer, Frederick (1860-1903), 49-56, 60, 83
Beer, Julius, 49, 50
Beer, Rachel (née Sassoon), 17, 18, 26, 29-32, 36, 39, 40, 46-56, 60, 81, 83, 99, 105, 107
Berkeley, Lady, 20
Berkeley, 7th Lord, 20
Berkshire Yeomanry, 92
Berlin, 38, 105
Berlin, Sir Isiah, 103
Bernhardt, Sarah, 33
Bevis Marks, 18, 27, 28
Blackpool Golf Club, 79
Boadeccia, 41
Boer War, 81, 108
Bombay, 12-15, 17, 25-27, 29, 30, 89, 90
Bombay, Bank of, 13, 27
Bombay Cotton Exchange, 13
Boulogne, 64
Boyle, Richard (see 1st Viscount Shannon)
Boyne, Battle of, 20
Brenchley, 42, 45
Brighton, 30, 38
Bridge Street, Walton, 34, 93, 94
Brompton Hospital, 49
Brooklands Motor Track, 83
Brown, Ford Madox, 40
Brown, Frederick, 69
Bruce, Edgar, 73, 74
Burchett, 74
Burhill Estate, 78
Burhill Golf Club, 31, 79, 83
Burhill House, 79
Burne-Jones, Sir Edward, 75
Burnley Golf Club, 80
Bushire, 10, 11

Cababe, Paul, 45
Calcutta, 10, 11
Cambridge, Duke of, 52
Cambridge University, Clare College, 46, 47
Canaletto, 55
Cauer, R. 57
7th Cavalry Brigade, 90
1st Cavalry Division, 91
Chancellor House, Tunbridge Wells, 55
Charing Cross, 54
Chesterfield Gardens, Mayfair, 50, 51, 54, 55
Chesterfield House, Mayfair, 50
Chiswick, 41
Christie & Co., 55, 56, 98
Churchfield Road, Walton, 32
Church Street, Walton, 101
Church Walk, Walton, 81
Claremont, Esher, 67
Clea Hall, Cumberland, 20
Cobham, 57, 70
Constable, John, 55
Corot, 55
Cottimore Crescent, Walton, 45
Craiglockhart, Edinburgh, 46
Cresta Run, 105
Crimean War, 37, 87
Cumberland, Ernest Duke of, 18
Cumberland Terrace, 19, 28
Curzon Street, Mayfair, 50
Cypress Lodge, 80

Damon, Rev. 'Hoppy', 68
Daud Pasha, 10
David, King, 9
David Sassoon & Sons, 12-14, 18, 25, 97
Davis, A.O., 104
Davis, Lionel, 35
Dawes, Dr., 71
Derby Lodge, Sunbury, 21
Devil's Island, 51
Dicey, William, 50
Disraeli, Benjamin, 52
5th Dragoon Guards, 73
Dreyfus, 51, 52
Drill Hall, Walton, 93
Dublin, 20

Eastbourne, 44
Eastern Mediterranean Squadron, 91
East India Buildings, 18
East India Company, 11-14, 18, 19
East India House, 18
E.D. Sassoon & Co., 28
Edward VI, King, 19
Edward, Prince of Wales (later King Edward VII), 28, 30, 38, 42, 75
Edward, Prince of Wales (later King Edward VIII), 18
Egyptian Expeditionary Force, 92
Elmbridge Borough Council, 104
Elphinstone, Lord, 14
Esterhazy, Major, 51, 52
Eton College, 62, 68-70, 80-83, 85, 88, 89, 106
Euphrates, River, 89
Eves, Mr, 32
Ezra, Muriel (née Sassoon), 97
Ezra ben Rahel, 10

Faure, M., President of France, 62
Fleet Street, 51
Fletcher, Constance (George Fleming), 43
Fletcher, Sir Henry, 1st Bart., 20
Fletcher, Sir Henry, 2nd Bart., 21
Fletcher, Sir Henry, 3rd Bart., 21
Fletcher, Sir Henry, 4th Bart., 19, 21, 22
Fletcher, John, 21
Forster, E.M., 45
Francis, Lazarus, 66, 67
Frankfurt, 49, 50
Franklin, Dulcie (see Sassoon, Dulcie)
Franklin, Frederick, 91, 99
Franklin, Margaret (see Sassoon, Margaret)
French Revolution, 74
Frere, Sir Bartle, 26
Friary, Holroyd Brewery Co., 77

Gallipoli, 46, 91, 92, 95
Galloway, 105
Gatty, Hester (see Sassoon, Hester)
Gawsworth, 40
General Strike 1926, 107
George, Prince of Wales (later King George III), 20
George IV, King, 18
Gibson, Mr Graeme, 101
Gibson, Mrs Graeme, 98, 101
Gladstone, Catherine, 26, 50
Gladstone, William E., 26-28, 50, 52, 57
Glasgow, 108
Golders Green Jewish Cemetery, 102
Grosse, Edmund, 40
Grosse, Nellie, 40
Grace, W.G., 31, 74, 75
Grand Parade, Eastbourne, 44
Grange over Sands Golf Club, 80
Gray Dawes & Co., 107
Great Exhibition, London, 1851, 40
Great Tidworth Farm, Gawsworth, 40
Great War, 39-41, 46, 55, 74, 80, 87, 108
Greest, Ben, 56
Grosvenor Square, London, 102
Guinness Family Trust, 78
Guinness, The Hon. Rupert, 78
Gunzburg, Baron Horace, 36, 38, 58, 64, 82, 108
Gunzburg, Baron Vladimir, 97
Gunzburg, Germany, 37
Gunzburg, Penelope (née Sassoon), 56, 100, 108
Gunzburg, Serge, 108

Haig, Sir Douglas (later Earl Haig), 89
Hampton Court Palace, 19
Handasyde, Elizabeth (née Thornycroft), 41
Handasyde, Major General Roger, 41
Harmsworth, Alfred (see Northcliffe, Lord)
Harwich & Dovercourt Golf Club, 80
Hawksley, B.F., 97
Heatherdene, Oatlands, 104
Henry VIII, King, 19, 22
Hepworth, Cecil, 93, 95

Heralds, College of, 19
Hersham, 84
Hersham Road, Walton, 34, 45, 75, 83
Hesse, Grand Duke of, 37
Heytesbury House, 47, 48, 56
Hicks, Joy, 105
Highgate Cemetery, 54
Hindhead, Beacon Hotel, 71, 85
Holy Trinity Church, Sloane Street, 50
Homebase Store, Walton, 104
Hong Kong, 13, 26
Honourable Artillery Company, 94, 95
Hoppner, 55
Hove, 35, 38, 39, 44, 49, 99
Howard, Miss, 55
Hurst Grove, Walton, 93
13th Hussars, 70, 85, 89, 90
Hyeem, Farha (see Sassoon, Farha)
Hythe, 89

I. Albury & Co., 107
Imperial War Museum, 105
7th Indian Cavalry Brigade, 90
2nd Indian Cavalry Division (Meerut Cavalry Brigade), 89
Indian Mutiny, 14
18th Infantry Division, 99, 106
6th Inniskilling Dragoon Guards, 68, 86, 87
Iveagh, Viscount (later Iveagh, Earl of), 78

Jacintha, 105
Jardine Matheson & Co., 63
Jerusalem, 92
Jews College, 26
Jews Free School, 26
John, Augustus, 69, 105
Jones, H. Thadeus, 47
Joseph, Hannah (see Sassoon, Hannah)
Journalists, Institute of, 51
Jubilee Place, Chelsea, 105
Justerini & Brooks, 60

Kensington Gore, 43
Key, Walter C., 80
Kindersley, R.S., 69
Kingston on Thames, 57
Kirkcudbright, 77, 105
Kipling, Rudyard, 49
Kut el Amera, 89, 90

30th Labour Corps (later Royal Pioneer Corps), 88
138th Labour Co., 88
Lancaster Gate, 30, 91
1st Lancers, 88
Langdon, Sir Godfrey, 93
Langham Hotel, 28
Langtry, Lily, 75
Leadenhall Street, 18
Leopold, Prince, Duke of Albany, 67
Lewis, Sir George, 1st Bart., 51, 75
Lewis, Katie, 75
Links, The, Walton, 34
Liverpool, 25, 26
Lloyds, Corporation of, 108
London Regiment, 96
London Stock Exchange, 49, 95
London University, Slade School of Art, 69, 82, 85
Lousada, Herbert, 46
Luker, George, 32
Lyric Cricket Club, 74
Lytham & St. Annes Golf Club (later Royal Lytham & St. Annes Golf Club), 79
Lytham Green Drive Golf Club, 80

Mabel (nursemaid), 62
Macclesfield, 40
Magnus, Philip, 99
Malabar Hills, 15
Manchester, 25, 26
Manchester Cotton Exchange, 27
Margaret, Princess, 104
Marlborough College, 46, 47
Marseilles, 89
Martinpuich, 91
Massay, Yvonne (see Sassoon, Yvonne)
Matfield, 42
Maud (nursemaid), 62
Meade, General, 90
Mersey, River, 46
Mesopotamia, 9, 13, 89, 90
Meyer, Doris (see Sassoon, Doris)
Meyer, Mr & Mrs Elias, 102
Middlesex, Lady, 20
Middlesex Regiment, 96
Mile End Jewish Cemetery, 28, 45, 71, 97, 101
Millais, Sir John, 55
Milton Abbas, 107
Mitchell, Mrs, 43-45
Molesey, 57
Mons, 88, 91
Mosul, 90
Mount Felix, 91, 94, 95, 99, 101, 104
Murat, Marechal, 107
Murdoch, W.L., 31, 73-75
Murray, Miss, 101, 102

Nanking, Treaty of, 13
Napoleon, 107
Nash, John, 18
Neuve Chapelle, 89
Neuve Eglise, 88
New Zealand Avenue, 104
New Zealand High Commissioner, 100
New Zealand Military Hospital, 95, 99, 104
New Zealand War Contingent Hospital, 94
Northcliffe, Lord, 52

Oates, Captain, 87
Oatlands, 84
Oatlands Chase, 34, 93, 103
Oatlands Drive, 34, 74, 75, 101, 103
Oatlands Palace, 19
Oatlands Park House (Hotel), 95
Oatlands Sailor's Home, 101
Observer Newspaper, 49-52, 81
O'Connor, T.P., MP, 53
Onslow Gardens, Chelsea, 69
Opium Wars, 13, 14
Owen, Wilfred, 46
Oxford University, 26, 30, 31, 40, 41, 46, 47, 82
Oxford University, Christ Church College, 31, 68, 82, 85
Oxford University, Exeter College, 33
Oxford University, Queens College, 30

Paddock Wood, 56
Palestine, 92
Paris, 33, 36-38, 40, 51, 85, 105
Parliament House, Dublin, 20
Parliament, Houses of, London, 41
Paul Graupe Gallery, Berlin, 105
Pearce, Sir Edward, 20
Pearson, Mr Justice, 34
Pendle Forest Golf Club, 80
Pembroke Gardens, South Kensington, 43
Persia, Shah of, 42, 43
Perugini, Edward, 29
Ploegsteert, 88
Poona, 26
Poona General Hospital, 14
Portland Place, 28, 49, 75
5th Princess Charlotte of Wales' Dragoon Guards, 71, 85, 91
Professional Golfers Association, 80
Punch Magazine, 53

2nd (Queens Bays) Dragoon Guards, 88
2nd Queens Royal West Surrey Regiment, 88

Read, Arthur, 30-33, 61-65, 70, 72-75, 77-81, 96, 97, 101-103
Read, Robert, 30
Reed, Walter W., 31, 73-75
Reed's School, Cobham, 70
Red Cross, 94
Reigate, 30
Reuben, Farha (see Sassoon, Farha)
Richmond, Surrey, 51, 53
Rivers, Dr. William, 46
Riviera, French, 50, 108
Rosenburg, Vera (see Sassoon, Vera)
Ross, Miss E.M., 47, 55, 56
Rothschild, Leopold de, 18
Roubilliac, Louis, 20
Royal Academy, 40-42
Royal & Ancient Golf Club, St. Andrews, 76
Royal Artillery, 70, 85, 90
Royal College of Surgeons, 69
Royal Flying Corps, 105
Royal Naval Air Service, 91
Royal Navy, 95
Royal Military Academy, Woolwich, 70, 85
Royal Military College, Sandhurst, 70, 71, 85
Royal Welch Fusiliers, 46
Rushworth, Dr. Norman, 97
Rusthall, Kent, 55, 56
Rydens Road, Walton, 96

St. Anne's Wells Gardens, Hove, 39
St. Mary Abbotts Church, Kensington, 41
St. Mary's Parish Church, Walton, 20, 21, 31, 32, 41, 59, 66, 83, 100, 103
St. Marys Road, Oatlands, 101
St. Paul's Church, Rusthall, 56
St. Petersburg, 36-38, 60
Salisbury, Marquess of, 52
Sandhurst (see Royal Military College)
Sandroyd Preparatory School, Cobham, 70, 83
Sargent, 75
Sason ben Salah (1750-1830), 9, 10
Sassoon, Aaron (1841-1907), 12
Sassoon, Abdullah (see Sassoon, Albert)
Sassoon, Abraham (see Sassoon, Arthur)
Sassoon, Albert (Abdullah) 1st Bart. (1818-1896), 10, 12-15, 17, 26, 27, 38, 42, 43
Sassoon, Alfred (1861-1895), 19, 29, 30, 32-34, 36, 40-46, 53, 60, 97

Sassoon, Arthur (Abraham) (1840-1912), 12, 17, 30, 35, 42
Sassoon, Arthur (1889-1960), 10, 11, 56, 61, 62, 68, 70, 82, 85, 89, 90, 99, 106
Sassoon, Beatrice, 107
Sassoon Cottages, Walton, 45
Sassoon, Cynthia, 107
Sassoon, David (1792-1864), 9, 10, 12-17, 19, 25-27, 36
Sassoon, David (1888-1978), 61, 62, 66, 67, 71, 77, 79, 81-83, 85, 87, 88, 92, 102, 105, 106, 108
Sassoon, David Solomon (1880-1942), 96
Sassoon Dock, Bombay, 89
Sassoon, Doris (née Meyer) (1896-1975), 102, 107
Sassoon, Dulcie (née Franklin), 91, 97, 99, 100, 102, 108
Sassoon, Edward, 2nd Bart. (1856-1912), 68, 97
Sassoon, Edward (Teddie) (1892-1955), 61, 62, 70, 71, 73, 77, 79, 80, 85, 91, 92, 108
Sassoon, Elias (1820-1880), 12-15, 17, 27-29, 31, 34
Sassoon, Farha (née Hyeem), 10, 30
Sassoon, Flora (Farha) (née Reuben) (1837-1919), 15, 17-19, 24-30, 32-38, 40, 41, 44, 45, 49, 50, 53, 58, 60, 97-99, 103
Sassoon, Frederick (1853-1917), 12
Sassoon, Frederick (1862-1889), 19, 29, 30, 32, 34-36, 38, 97
Sassoon, Frederick (Freddie) (1892-1979), 61, 62, 70, 71, 77, 79, 80, 85, 90, 91, 95, 99, 107, 108
Sassoon, (Gabriel) Jacques (1930-), 106
Sassoon, George (1930-), 47
Sassoon, Hamo (1887-1915), 42, 44-47
Sassoon, Hannah (née Joseph) (1794-1826), 10
Sassoon, Hannah (Missy) (1887-1910), 61, 62, 71, 83-85, 97
Sassoon, Hester (née Gatty), 47
Sassoon, Hugh, 107
Sassoon, John, 70, 107
Sassoon, Joseph (1855-1918), 11, 15, 18, 29-38, 41, 44, 45, 47, 53, 55, 57-60, 62-68, 70, 72-86, 92, 93, 95-97, 99 101-104, 108
Sassoon, Joseph (1927-), 106
Sassoon, Louise (née Gunzburg) (1862-1921), 36-38, 57, 60, 62-65, 67, 69, 72, 76, 77, 80-83, 86, 92-103, 105, 108
Sassoon, Margaret (née Franklin) (1887-1952), 91, 95, 99, 107, 108
Sassoon Mechanics Institute, 14, 26
Sassoon, Michael (1884-1969), 42, 44, 45, 56
Sassoon, Minnie (née Channing), 108
Sassoon, Mozelle (see Spielman, Mozelle)
Sassoon, Muriel (see Ezra, Muriel)
Sassoon, Penelope (see Gunzburg, Penelope)
Sassoon, Philip 3rd. Bart., (1888-1939), 68, 70, 89
Sassoon, Rachel (see Beer, Rachel)
Sassoon, Reuben (1835-1905), 12, 15, 17, 28, 30, 33, 38, 66
Sassoon, Richard, 70 107
Sassoon, Sassoon David (S.D.), 11, 13, 15-19, 21, 23-31, 33, 49, 58, 97
Sassoon, Sassoon Joseph (Totts), (1885-1922), 56, 60-62, 68-70, 73, 77, 81, 82, 85, 87, 88, 91, 93, 94, 97, 99-103, 108
Sassoon, Seigfried (1886-1967), 40, 43-47, 50, 52-54, 56
Sassoon, Solomon (1841-1894), 12, 29, 34, 35
Sassoon, Theresa (née Thornycroft), 36, 40-43, 45, 47, 50-56, 60
Sassoon, Vera (née Rosenburg), 82, 83, 105, 106, 108
Sassoon, Yvonne (née Massay), 107
Scott, Captain Robert F., 87
Sefi, Antonio, 29, 34, 35
Sevenoaks, 107
Shanghai,13-16, 19, 26, 29
Shannon, Lady, 20
Shannon, 1st Viscount, 20, 21, 23, 72
Shatt al Hai, River, 90
Shoshan, Ibn, 9
Sidney Road, Walton, 45
Simpson, Joseph, 105
Simpson, Wallis (later Duchess of Windsor), 18
Slade, Felix, 69
Slade School of Art (see London University)
Somme, River, 46, 88, 91
South Pole, 87
Spanish & Portuguese Synagogue, 18, 27, 28
Spencer, Stanley, 69
Spielman, Brian, 107
Spielman, Edgar (1885-1978), 95, 97, 107
Spielman, Hannah, 107
Spielman, Mozelle (née Sassoon) (1890-1973), 61, 62, 71, 84, 85, 87, 95, 97, 102, 107
Spielmam, Nancy, 107
Spofforth, F.R., 75, 100
Springfield, Rydens Road, 96, 97, 101
Station Avenue, Walton, 34, 67, 75, 83, 103
Steer, Edward, 80
Steer, Gordon, 79
Steer, Herbert, 79
Steer, James Albert, 76, 80
Steer, James Junior, 76, 79
Steer, Joseph, 80
Stevens, 'Lumpy', 101
Stock Exchange, London, 62
Stompond Lane (formerly Stompond Road), 34, 75, 83, 84, 104
Stowe School, 70
Suleiman, Pasha, 10
Sunday Special Newspaper, 53
Sunday Times Newspaper, 52, 53, 81
Surbiton, 100
Surrey County Cricket Club, 31, 74
Surrey Herald Newspaper, 83
Surrey Staghounds, 80
Sussex Yeomanry, 46
Sutherland, Graham, 69

Tamarind Lane, Bombay, 11
Tankerville, Earl of, 101
Taylor, Alma, 95
Terrace Road, Walton, 45
Thames Ditton, 57
Thames, River, 23
Theipval, 89
Thornycroft, Agatha, 40
Thornycroft Hall, Cheshire, 41
Thornycroft, Hamo, 40, 41, 46
Thornycroft, Helen, 50
Thornycroft, Sir John Bart., 41
Thornycroft, John, 40, 41
Thornycroft, Mary, 40
Thornycroft, Theresa (Georgiana) (see Sassoon, Theresa)
Thornycroft, Thomas, 40, 41
Tigris, River, 89
Teledo, Spain, 9
Tonks, Henry, 69
Trevelyan, G.M., 74
Trigg's Dairy, 81
Trigg, Jesse, 81
Trigg, Jesse, Jnr., 81, 86
Tunbridge Wells, 42, 55, 60, 99

Ulm, 37
United Dairies, 81

Volunteer Training Corps, 93
Victoria & Albert Museum,18, 28
Victoria Art Gallery, Melbourne, 12
Victoria, Queen, 12, 40, 67, 68

Wagner, 42
Walton & Weybridge Urban District Council, 104
Walton Bridge, 23, 101
Walton Cottage Hospital, 103
Walton Cricket Club, 30, 74, 75, 100
Walton Enclosure Award, 21
Walton Grove, 45
Walton High Street, 24, 32, 34, 81, 86, 101, 104
Walton Lodge, 41
Walton Public Hall, 32, 57
Walton Regatta, 68, 73
Walton Station (formerly Walton & Hersham Station), 83
Walton Town Hall, 104
Walton Urban District Council, 84, 99, 101, 104
Walton Women's Club, 86
Warminster, Wiltshire, 47
Warner, George, 32
War Office, Directorate of Military Intelligence, 88, 94
War Supply Depot, Walton, 87
Waterers, 83, 98, 102
Watford, 56
Weald of Kent, 42
Weirleigh, 42, 43, 46, 47, 52, 53, 56
Wellington, Dowager Duchess of, 78
West Africa, 107
West End Synagogue, Bayswater, 97
Western Isles, Scotland, 47
Westminster Bridge, 41
Westminster Volunteer Rifles, 94
Weybridge, 45, 57
Wilson (headmaster), 70
Windsor Castle, 68
Wingrove, Gus, 53
Wisden & Co., 52
Woodcote, Ashley Road, 93
Wood, Lady Mary, 28
Woolner, Thomas, R.A., 27, 28
Woolwich (see Royal Military Academy)

World War II, 70

Ypres, 88, 91